THE TEACHER'S GUIDE

to

THE METRIC SYSTEM

THE TEACHER'S GUIDE
to
THE METRIC SYSTEM

Compiled and Edited by

A. L. Le MARAIC

and

J. P. CIARAMELLA

PUBLISHED BY ABBEY BOOKS

SOMERS, N.Y. 10589

ISBN Prefix 0–913768

THE TEACHER'S GUIDE TO THE METRIC SYSTEM

First Edition
L. C. Card Number 74–3811
ISBN Number 0–913768–01–4

Printed in the United States of America.

PREFACE

This compilation is intended as a guide to the proper use of the new international metric system known as Le Système International d'Unités (SI), and to give in simplified form its underlying and unchanging principles so as to aid teachers and others to put it to practical use as soon as possible in keeping with the aims and efforts of the United States Government and the international community of nations.

Teachers at all educational levels are placed in the unique and important position of rendering a service of inestimable value by disseminating its basic principles and helping to establish its correct usage during the difficult period of transition when we must change from old methods of measurement to the newly adopted universal system.

The International System of Units (SI) is a new concept updating the traditional metric system of the past. Its newness, however, does not represent a detachment from the old system, rather, a continuity establishing simplified, consistent, and coherent standards of metric weights and measures to be used by all countries of the world henceforth.

We hope this book will serve as an aid to teachers at all educational levels in planning and developing progressive work programs towards the enhancement of metric skills in schools.

OTHER METRIC BOOKS

THE COMPLETE METRIC SYSTEM with
THE INTERNATIONAL SYSTEM OF UNITS (SI)
L. C. Card Number 72–97799   ISBN Number 0–913768–00–6

THE METRIC SYSTEM FOR BEGINNERS
L. C. Card Number 74–3812   ISBN Number 0–913768–02–2

THE METRIC SYSTEM FOR SECONDARY SCHOOLS
L. C. Card Number 74–10176   ISBN Number 0–913768–04–9

THE METRIC ENCYCLOPEDIA
L. C. Card Number 74–9235   ISBN Number 0–913768–03–0

(Send for our list of metric classroom charts and posters.)

CONTENTS

ACKNOWLEDGMENT

All the basic material used in this book is derived from the translation approved by the International Bureau of Weights and Measures of its publication "Le Système International d'Unités." Grateful acknowledgment is made to the following for permission to include excerpts from their publications in this compendium.

Conférence Générale des Poids et Mesures   (CGPM)
Consultative Committee for Units   (CCU)
International Bureau of Weights and Measures   (BIPM)
International Committee of Weights and Measures   (CIPM)
International Organization for Standardization   (ISO)
National Aeronautics and Space Administration   (NASA)
National Bureau of Standards—U.S.A.   (NBS)
South African Bureau of Standards   (SABS)
South African Metrication Advisory Board   (SAMAB)
United States Department of Commerce   (USDC)

INTRODUCTION

It has taken almost four centuries to bring the metric system to the point of universal acceptance by all nations of the world as the only legal system of measurements to be used henceforth. Almost four centuries ago, in 1585, the Flemish mathematician, Simon Stevin, conceived the idea of using a decimal system of measurement and advocated its adoption with no success. In the following century (17th) astronomers, mathematicians, and scientists made similar proposals regarding the use of decimal units of measurement but they too failed to arouse much interest. Later on, during the 18th century, when the French Revolution brought about many radical reforms, the idea was revived again and in 1790 the French decided to take definitive steps to establish a universal system of measurement based on decimals as originally conceived by Stevin two centuries earlier. As a result, in 1793, the "*metre*" was proposed as the unit of length, and the "*gramme*" as the unit of mass. Antoine Laurent Lavoisier, the French chemist and physicist hailed as the founder of modern chemistry, was one of the sponsors who spearheaded the movement for the adoption of the metric system but unfortunately he was guillotined before he could see it accomplished. Nevertheless, the new system continued to make progress in France and it soon spread to other European countries. It was not until nearly a century later (1875) that the Treaty of the Metre was signed by the United States and sixteen other nations, and in 1876 an International Bureau of Weights and Measures was founded in Sevres, France.

Although the metric system was based on sound principles of judgment, it gradually developed variations in different countries. These variations from country to country gave rise to many problems and disputes with the result that in 1960 a conference was called to establish a new, uniform system of standardized units of measure for universal use. This is what is now known as Le Système International d'Unités (SI), The International System of

Units. It has been accepted by international agreement as the only legal system of metric measurements to be used henceforth all over the world. The rapid expansion of trade between nations during the last two decades, the enormous strides made in industrial, scientific, and outer-space technology, and the ever increasing exchange of industrial and scientific techniques between nations have further emphasized the importance of having a standardized universal system of metric measurements.

July 1974   A. L. LeMaraic
                                                    J. P. Ciaramella

THE TEACHER'S GUIDE

to

THE METRIC SYSTEM

I. RELATIONSHIP BETWEEN THE METRIC SYSTEM AND THE DECIMAL SYSTEM

The metric system is based on the decimal system exactly like our United States monetary system of dollars and cents and it is just as easy to use. The only difference is that instead of dollars metric units of measures are substituted. There are seven base units in the new SI metric system but for practical purposes we will concentrate on the three most important ones for the time being, the *"metre"* for length; the *"gram"* for mass; and the *"litre"* for capacity.

Here is a simple illustration of the similarity between the metric system and our monetary system of dollars and cents:

U.S. Dollars	metre (m)	gram (g)	litre (l)
1.00	1.00 m	1.00 g	1.00 l
10 dimes (tenths)	10 dm	10 dg	10 dl
100 cents (hundredths)	100 cm	100 cg	100 cl
1,000 mils (thousandths)	1,000 mm	1,000 mg	1,000 ml

It must be noted that all the above are in some multiple of 10; 100; or 1,000.

The above metric values are also written in decimals exactly like United States dollars and cents.

$1.00	1.00 m	1.00 g	1.00 l
$0.10	0.10 m	0.10 g	0.10 l
$0.01	0.01 m	0.01 g	0.01 l
$0.001	0.001 m	0.001 g	0.001 l

Mixed metric decimal numbers expressing whole units and fractions of units are also written like United States dollars and cents.

3

U.S. Dollars	metres (m)	grams (g)	litres (l)
$1000.00	1,000 m	1,000 g	1,000 l
$ 100.00	100.00 m	100.00 g	100.00 l
$ 10.00	10.00 m	10.00 g	10.00 l
$ 1.00	1.00 m	1.00 g	1.00 l
$ 1.99	1.99 m	1.99 g	1.99 l
$ 1.75	1.75 m	1.75 g	1.75 l
$ 1.50	1.50 m	1.50 g	1.50 l
$ 1.25	1.25 m	1.25 g	1.25 l
$ 1.10	1.10 m	1.10 g	1.10 l
$ 1.05	1.05 m	1.05 g	1.05 l
$ 1.01	1.01 m	1.01 g	1.01 l
$ 0.99	0.99 m	0.99 g	0.99 l
$ 0.75	0.75 m	0.75 g	0.75 l
$ 0.50	0.50 m	0.50 g	0.50 l
$ 0.25	0.25 m	0.25 g	0.25 l
$ 0.10	0.10 m	0.10 g	0.10 l
$ 0.05	0.05 m	0.05 g	0.05 l
$ 0.01 (1 cent)	0.01 m	0.01 g	0.01 l
$ 0.001 (1 mil)	0.001 m	0.001 g	0.001 l

Here is another illustration showing how metric values are expressed the same as U.S. dollar values:

1 dollar, 75 cents; 1 metre, 75 centimetres; 1 gram, 75 centigrams; 1 litre, 75 centilitres.

1 dollar, 50 cents; 1 metre, 50 centimetres; 1 gram, 50 centigrams; 1 litre, 50 centilitres.

1 dollar, 25 cents; 1 metre, 25 centimetres; 1 gram, 25 centigrams; 1 litre, 25 centilitres.

1 dollar, 10 cents; 1 metre, 10 centimetres; 1 gram, 10 centigrams; 1 litre, 10 centilitres.

1 dollar, 5 cents; 1 metre, 5 centimetres; 1 gram, 5 centigrams; 1 litre, 5 centilitres.

1 dollar, 1 cent; 1 metre, 1 centimetre; 1 gram, 1 centigram; 1 litre, 1 centilitre.

It is to be noted from the above examples that the prefixes deci-; centi-; and milli- are commonly used with all metric base units.

All prefixes remain the same to express quantities greater than base units (*multiples*) or quantities smaller than base units (*submultiples*).

The Greek prefixes *deka-*, before a unit means ten; *hecto-*, one hundred; *kilo-*, one thousand; *myria-*, ten thousand; *mega-*, one million; thus *hectometre* means 100 metres.

The Latin prefixes *deci-*, before a unit means one tenth; *centi-*, one hundredth; *milli-*, one thousandth; *micro-*, one millionth; thus a *centigram* is one hundredth of a gram.

All the above prefixes are used commonly with all metric base units, i.e. metre, gram, litre, etc.

DECIMAL FRACTIONS

Here are some examples of decimal fractions written in expanded form.

$$0.1 = 1/10 \qquad \text{may be written } 1/10^1 \text{ or } 10^{-1}$$
$$0.01 = 1/100 \qquad \text{may be written } 1/10^2 \text{ or } 10^{-2}$$
$$0.001 = 1/1,000 \qquad \text{may be written } 1/10^3 \text{ or } 10^{-3}$$
$$0.0001 = 1/10,000 \text{ may be written } 1/10^4 \text{ or } 10^{-4}$$

The value of a decimal fraction may be expressed by a power of ten with a negative exponent (minus).

Let us break down the decimal fraction 0.7875

$$0.7875 = 7 \text{ tenths } + \; 8 \text{ hundredths } + \; 7 \text{ thousandths} + \; 5 \text{ ten-thousandths}$$

or

$$0.7875 = 7 \times 1/10 + 8 \times 1/100 + 7 \times 1/1,000 + 5 \times 1/10,000$$

or

$$0.7875 = 7 \times 1/10^1 + 8 \times 1/10^2 + 7 \times 1/10^3 + 5 \times 1/10^4$$

or

$$0.7875 = (7 \times 10^{-1}) + (8 \times 10^{-2}) + (7 \times 10^{-3})$$
$$+ (5 \times 10^{-4})$$

The mixed decimal number (whole number and fraction) 745.787 expresses the following:

$$(7 \times 10^2) + (4 \times 10^1) + (5 \times 10^0) +$$
$$(7 \times 10^{-1}) + (8 \times 10^{-2}) + (7 \times 10^{-3})$$

It is to be noted that place values of fractional powers of ten are in a decreasing order from left to right of the decimal sign while those for whole numbers increase from right to left of the decimal sign.

MORE ON DECIMAL NUMBERS AND FRACTIONS

The metric system is a decimal system and decimal fractions will therefore replace ordinary fractions. You will therefore have to become conversant with decimal numbers and how they are written.

The place values in a number, viz. 3 768,452; may be represented as follows:

Thousands	Hundreds	Tens	Units
3	7	6	8

Tenths	Hundredths	Thousands
4	5	2

If you have to add two numbers, it is clear that tens may be added to tens only or hundredths to hundredths. If you wish to add 31,5 and 3,25 you will go about it as follows:

$$\begin{array}{r} 31,5 \\ + \quad 3,25 \\ \hline 34,75 \end{array}$$

If you wish to ascertain what ordinary fraction is represented by a specific decimal fraction, you will again have to take the place values into consideration. Take 0,5 for instance: from the above it is apparent that 0,5 represents 0 units + 5 tenths, i.e. $\frac{5}{10}$ which equals $\frac{1}{2}$. Similarly 2,75 equals 2 units + 7 tenths + 5 hundredths or $2 + \frac{7}{10} + \frac{5}{100}$. Since $\frac{7}{10} = \frac{70}{100}$, we find:

$$2 \quad + \quad \frac{70 + 5}{100}$$

$$= \quad 2 \quad + \quad \frac{75}{100}$$

$$= \quad 2 \quad + \quad \frac{3}{4}$$

$$= \quad 2\frac{3}{4}$$

If you wish to determine how many grams are contained in say 0,5 kg, you may set about it in two ways:

Since the prefix kilo- (k) means 1 000, you may conclude that there are 1 000 g in 1 kg. Therefore

$$
\begin{aligned}
&\quad\ 0,5 \quad \text{kg} \\
&= \quad \tfrac{5}{10} \quad \text{kg} \\
&= \quad \tfrac{1}{2} \quad \text{kg} \\
&= \quad 500 \quad \text{g}
\end{aligned}
$$

or   0,5   kg

= 0,500   kg (the quantity is not changed by adding noughts after the decimal sign)

= 0,500 $\times$ 1 000 g (there are 1 000 g in 1 kg)

= 500   g (when decimal numbers are multiplied by 1 000 the decimal sign is moved three places to the right)

You will soon know the most commonly used decimal fractions, viz:

$$0,1 \ = \ \tfrac{1}{10} \qquad\qquad\qquad 0,25 \ = \ \tfrac{1}{4}$$
$$0,2 \ = \ \tfrac{2}{10} \ = \ \tfrac{1}{5} \qquad\ \ 0,5 \ = \ \tfrac{1}{2}$$
$$0,3 \ = \ \tfrac{3}{10} \qquad\qquad\qquad 0,75 \ = \ \tfrac{3}{4}$$
$$0,4 \ = \ \tfrac{4}{10} \ = \ \tfrac{2}{5} \qquad\ \ 0,125 \ = \ \tfrac{1}{8}$$
$$0,5 \ = \ \tfrac{5}{10} \ = \ \tfrac{1}{2} \qquad\ \ 0,375 \ = \ \tfrac{3}{8}$$
$$0,6 \ = \ \tfrac{6}{10} \ = \ \tfrac{3}{5} \qquad\ \ 0,625 \ = \ \tfrac{5}{8}$$
$$0,7 \ = \ \tfrac{7}{10} \qquad\qquad\qquad 0,333 \ = \ \tfrac{1}{3}$$
$$0,8 \ = \ \tfrac{8}{10} \ = \ \tfrac{4}{5} \qquad\ \ 0,666 \ = \ \tfrac{2}{3}$$
$$0,9 \ = \ \tfrac{9}{10}$$

Remember:　　A zero is always written before the decimal sign in numbers less than 1, e.g. 0,25 and not ,25; etc.

Examples:

a) To multiply a number by 10; 100; or 1 000; the decimal sign is moved 1; 2; or 3 places to the right, depending on the number of noughts in the multiplier, e.g.

$$3,25 \times 10 = 32,5$$
$$3,25 \times 100 = 325$$
$$3,25 \times 1\ 000 = 3\ 250$$

Vice versa, to divide a number by 10; 100; or 1 000, the decimal sign is moved to the left, e.g.

$$112,4 \div 10 = 11,24$$
$$112,4 \div 100 = 1,124$$
$$112,4 \div 1\ 000 = 0,1124$$

b) Calculations involving decimal fractions are considerably easier than those involving ordinary fractions, so that much time and effort are saved. Let's assume that a housewife needs four panels of material for the curtains of a window.

Using a metric measuring tape, the sum will be as follows:

Length of one panel plus seam　　　:　　310　cm

Material required for four panels　　:　　310　cm
　　　　　　　　　　　　　　　　　　　　$\times$ 4
　　　　　　　　　　　　　　　　　　1 240　cm

What length of material will she have to buy?

Divide by 100　　　　　:　　100 ⌊ 1 240　cm
(100　cm = 1　m)　　　　　　　　　　12,40　m

(The decimal sign is moved two places to the left.)

She will therefore need 12,4 m of material for the curtains.

EXPONENTS AND WHAT THEY MEAN

When we multiply equal factors such as 5×5, 6×6, 7×7, they may be written 5^2, 6^2, 7^2 meaning 5 squared, 6 squared, 7 squared, etc.

When we use equal factors in multiplication to the 3rd power, i.e. $5 \times 5 \times 5$, $6 \times 6 \times 6$, $7 \times 7 \times 7$, they may be written 5^3, 6^3, 7^3 meaning the cube of 5, 6, 7, etc.

When we use equal factors in multiplication to the 4th power, i.e. $5 \times 5 \times 5 \times 5$, $6 \times 6 \times 6 \times 6$, $7 \times 7 \times 7 \times 7$, they may be written 5^4, 6^4, 7^4 meaning 5, 6, 7 to the 4th power, etc.

When we use equal factors in multiplication to the 5th power, i.e. $5 \times 5 \times 5 \times 5 \times 5$, $6 \times 6 \times 6 \times 6 \times 6$, $7 \times 7 \times 7 \times 7 \times 7$, they may be written 5^5, 6^5, 7^5 meaning 5, 6, 7 to the 5th power, etc.

The small figure at the upper right (the exponent) indicates how many times the same factor is to be used in multiplication, viz. to the 2nd, 3rd, 4th, 5th, 6th, 7th power, etc., as shown above.

It must be noted that 10^1, 10^2, 10^3, 10^4, 10^5, 10^6, 10^7, 10^8, 10^9, are all powers of TEN.

$$10^1 = 10 \times 1 = 10$$
$$10^2 = 10 \times 10 = 100$$
$$10^3 = 10 \times 10 \times 10 = 1,000$$
$$10^4 = 10 \times 10 \times 10 \times 10 = 10,000$$
$$10^5 = 10 \times 10 \times 10 \times 10 \times 10 = 100,000$$
$$10^6 = 10 \times 10 \times 10 \times 10 \times 10 \times 10 = 1,000,000, \text{ etc.}$$

Particular attention is called to the above exponents as they will be used quite often in making metric calculations.

II. WHAT THE METRIC SYSTEM IS ALL ABOUT

In its original conception the metre was the fundamental unit of the metric system, and all units of length and capacity were to be derived directly from the metre which was intended to be equal to one ten-millionth of the earth's quadrant. Furthermore, it was originally planned that the unit of mass, the kilogram, should be identical with the mass of a cubic decimetre of water at its maximum density. At present, however, the units of length and mass are defined independently of these conceptions.

A new definition of the metre in terms of the wave length of light was adopted by the 11th General (International) Conference on Weights and Measures, in 1960. According to this definition, 1 metre = 1 650 763.73 wave lengths in a vacuum of orange-red radiation of krypton 86.

The kilogram is independently defined as the mass of a definite platinum-iridium standard, the International Prototype Kilogram, which is also kept at the International Bureau of Weights and Measures. The litre is defined as the volume of a kilogram of water, at standard atmospheric pressure, and at the temperature of its maximum density, approximately 4° C. The metre is thus the fundamental unit on which are based all metric standards and measurements of length and area, and of volumes derived from linear measurements.

The kilogram is the fundamental unit on which are based all metric standards of mass. The litre is a secondary or derived unit of capacity or volume. The litre is larger by about 27 parts per million than the cube of the tenth of the metre, i.e., the cubic decimetre—that is, 1 litre = 1.000 027 cubic decimetres.

All lengths, areas, and cubic measures in the following tables are derived from the international metre, the basic relation between units of the customary and the metric systems being:

$$1 \text{ metre} = 39.37 \text{ inches,}$$

contained in the act of Congress of 1866. From this relation it follows that 1 inch = 25.400 05 millimetres (nearly).

All capacities are based on the equivalent 1 litre equals 1.000 027 cubic decimetres. The decimetre is equal to 3.937

inches in accordance with the legal equivalent of the metre given above. The gallon referred to in the tables is the United States gallon of 231 cubic inches. The bushel is the United States bushel of 2 150.42 cubic inches. These units must not be confused with the British units of the same name which differ from those used in the United States. The British gallon (277.420 cubic inches) is approximately 20 percent larger, and the British bushel (2 219.36 cubic inches) is 3 percent larger than the corresponding units used in this country.

All weights are derived from the International Kilogram, as authorized. The relation used is 1 avoirdupois pound = 453.592 427 7 grams.

In the construction of the tables in this compendium, when the fundamental relation of the units furnished directly a reduction factor for use in determining the multiples of the units, this factor was used in its fundamental form, as, for example, that 1 metre = 39.37 inches. Reduction factors which it was necessary to obtain, however, by multiplication, division, powers, or roots, etc., of the fundamental relations were usually carried out to a greater degree of accuracy than that to which it is usually possible to make measurements, for convenience in computing the multiples to the accuracy desired.

When the tables do not give the equivalent of any desired quantity directly and completely, the equivalent can usually be obtained, without the necessity of making a multiplication of these reduction factors, by using quantities from several tables, making a shift of decimal points, if necessary, and merely adding the results.

The supplementary metric units are formed by combining the words "metre"; "gram," and "litre" with numerical prefixes, as in the following table:

Prefixes		Meaning			Units
milli-	=	one-thousandth	$\frac{1}{1000}$	0.001	
centi-	=	one-hundredth	$\frac{1}{100}$	0.01	"metre" *for length.*
deci-	=	one-tenth	$\frac{1}{10}$	0.1	
Unit =		one		1	"gram" *for weight or mass.*
deka-	=	ten	$\frac{10}{1}$	10	
hecto-	=	one hundred	$\frac{100}{1}$	100	"litre" *for capacity.*
kilo-	=	one thousand	$\frac{1000}{1}$	1000	

DEFINITIONS OF UNITS

1. Length

Fundamental Units

A metre (m) is a unit of length equal to 1 650 763.73 wave lengths in a vacuum of orange-red radiation of krypton 86.

Multiples and Submultiples

1 kilometre (km) = 1 000 metres.
1 hectometre (hm) = 100 metres.
1 dekametre (dkm) = 10 metres.
1 decimetre (dm) = 0.1 metre.
1 centimetre (cm) = 0.01 metre.
1 millimetre (mm) = 0.001 metre.
1 micron (μ) = 0.000 001 metre = 0.001 millimetre.
1 millimicron (mμ) = 0.000 000 001 metre
                    = 0.001 micron.
1 angstrom (A) $\begin{cases} = 0.000\ 000\ 1 \text{ millimetre.} \\ = 0.000\ 1 \quad\quad \text{micron.} \\ = 0.1 \quad\quad\quad\ \text{millimicron.} \end{cases}$

2. Area

Fundamental Units

A square metre (m²) is a unit of area equal to the area of a square the sides of which are 1 metre.

Multiples and Submultiples

1 square kilometre (km²) = 1 000 000 square metres.
1 hectare (ha), or square hectometre (hm²)
     = 10 000 square metres.
1 are (a), or square dekametre (dkm²)
     = 100 square metres.
1 centare (ca) = 1 square metre.
1 square decimetre (dm²) = 0.01 square metre.
1 square centimetre (cm²) = 0.000 1 square metre.
1 square millimetre (mm²) = 0.000 001 square metre.

3. Volume

Fundamental Units

A cubic metre (m³) is a unit of volume equal to a cube the edges of which are 1 metre.

Multiples and Submultiples

1 cubic kilometre (km³) = 1 000 000 000 cubic metres.
1 cubic hectometre (hm³) = 1 000 000 cubic metres.
1 cubic dekametre (dkm³) = 1 000 cubic metres.
1 stere (s) = 1 cubic metre.
1 cubic decimetre (dm³) = 0.001 cubic metre.
1 cubic centimetre (cm³) = 0.000 001 cubic metre
                             = 0.001 cubic decimetre.
1 cubic millimetre (mm³) = 0.000 000 001 cubic metre
                             = 0.001 cubic centimetre.

4. Capacity

Fundamental Units

A litre (*l*) is a unit of capacity equal to the volume occupied by the mass of 1 kilogram of pure water at its maximum density (at a temperature of 4° C, practically) and under the standard atmospheric pressure (of 760 mm). It is equivalent in volume to 1.000 027 cubic decimetres.

Multiples and Submultiples

1 hectolitre (h*l*) = 100 litres.
1 dekalitre (dk*l*) = 10 litres.
1 decilitre (d*l*) = 0.1 litre.
1 centilitre (c*l*) = 0.01 litre.
1 millilitre (m*l*) = 0.001 litre
                    = 1.000 027 cubic centimetres.

5. Mass

Fundamental Units

A kilogram (kg) is a unit of mass equal to the mass of the International Prototype Kilogram.

A gram (g) is a unit of mass equal to one-thousandth of the mass of the International Protoype Kilogram.

Multiples and Submultiples

1 metric ton (t) = 1 000 kilograms.
1 hectogram (hg) = 100 grams.
1 dekagram (dkg) = 10 grams.
1 decigram (dg) = 0.1 gram.
1 centigram (cg) = 0.01 gram.
1 milligram (mg) = 0.001 gram.
1 metric carat (c) = 200 milligrams = 0.2 gram.

STANDARDS OF MEASUREMENT

Units of measurement should be distinguished from standards of measurement. Units of length are fixed distances, independent of any other consideration, while length standards are affected by the expansion and contraction resulting from changes of temperature of the material of which the standard is composed. It is therefore necessary to fix upon some temperature at which the distance between the defining lines or end surfaces of the standards shall be equal to the unit. The same is true of standards of capacity, which at some definite temperature contain a given number of units of volume.

The recommended standard temperature for commercial and industrial length standards is 20° C (68° F). Some metric standards, especially those made in Europe until recently, are intended to be correct at 0° C. In the past some length standards graduated in the customary units have been made to be correct at 62° F (16.67° C).

For measurements of high precision it is also necessary to specify the manner of support of the standards, whether at certain points only or throughout their entire length, and in the case of tapes it is also necessary to give the tension applied to the tape when in use.

In the United States the capacity standards, both metric and customary, are made to hold the specified volumes at 4° C. Standards of capacity are usually made of brass and the capacity at any other temperature may be computed by the use of the coefficient of cubical expansion of that material usually assumed to be 0.000 054 per degree centigrade. In the purchase and sale of liquids a more important consideration than the temperature of the measures is the temperature of the liquid when measured, for the reason that the large coefficient of expansion of many liquids makes the actual mass of a given volume delivered vary considerably with temperature.

While the temperature of a masspiece does not affect its mass, it is nevertheless important that when two masspieces are compared in air they both be at the same temperature as the air. If there is a difference between the temperature of the air and the masspieces, convection currents will be set up and the readings of the balance will be thereby affected. Also, since masspieces are buoyed up by the surrounding air by amounts dependent upon their volumes, it is desirable that the masspieces of any set

14

be of the same material. If two masspieces of the same density balance in air of a certain density they will balance in vacuo or in air of a different density.

Brass is the material most widely used for standard masspieces, although platinum and aluminum are quite commonly used for masspieces of 1 gram or less. In the absence of any knowledge as to the actual density of masspieces, those made of brass are assumed to have a density of 8.4 grams per cm^3 at 0° C, while those of platinum and of aluminum are assumed to have densities of 21.5 and 2.7 grams per cm^3, respectively.

III. SIMPLICITY OF THE METRIC SYSTEM

One of the greatest advantages of the metric system is its simplicity. By learning the meaning of only a few terms, enough knowledge of the metric system can be gained to satisfy everyday needs. In the imperial system a wide variety of names had to be learned. The units of length you had to memorize were, among others, mile, yard, foot, and inch. Nothing in these words indicates that they all are units of length, or what their mutual ratio is. In the metric system, on the other hand, every unit of length has "metre" as part of the word and a suitable prefix indicates which multiple or submultiple of a metre is meant. Likewise, in the Imperial system the volume of liquids is measured in units such as the gallon, quart, pint, gill and fluid ounce. In the metric system the litre with its multiples and submultiples is used. It is therefore apparent that you need not be alarmed by the metric system.

The metric terms mentioned above can be divided into three groups:

Five prefixes: mega-
                   kilo-
                   deci-
                   centi-
                   milli-

Of these actually only three, kilo-, centi- and milli-, are of general importance.

Three main units: metre   (length)
                     kilogram   (mass)
                     litre   (volume)

Four further units: degree celsius   (temperature)
                     hectare   (area of farms)
                     metric ton   (mass)
                     bar   (pressure)

To simplify the change-over for yourself set a little time apart to learn the meaning of the prefixes and try to form an idea of the sizes of the main units. You will soon discover how easy calculations are in the metric system. Conversion factors need not be memorized, as a matter of fact, the continual conversion of metric to non-metric units will complicate the change-over considerably. Rather try to familiarize yourself with the actual sizes and quantities of the new units. Soft drinks, for instance, are already obtainable in litre bottles. Look at cans or bottles, pour the contents into a number of glasses, try to learn how much liquid is contained in a litre, and then forget the pint. As soon as milk products are sold in containers of 1 litre, 500 m*l* and 250 m*l* the housewife can calculate or guess how much she will need. For a few days she may perhaps buy too much or too little, but within a week she will have established a new pattern for her buying and soon she will have forgotten the quart and pint completely. Should the need arise for conversion, however, the factor should be used once only. The metric quantity can then be written next to the old designation after which it should be used always.

When next you buy a measuring tape see to it that it is a metric measuring tape. The same applies to a new thermometer or a kitchen or bathroom scale.

The metric terms mentioned above are discussed in more detail below.

UNITS FOR EVERYDAY USE

The SI is characterized by its simplicity. The particulars given below are sufficient to satisfy the everyday needs.

Metric Prefixes. Similar units are in decimal relation to one another. For all units, the same Greek prefixes indicate whether such unit is 10; 100; 1 000 or 1 000 000 times greater than the basic unit, and in the same way, Latin prefixes indicate whether the unit is $\frac{1}{10}$; $\frac{1}{100}$ or $\frac{1}{1\,000}$ of the basic unit.

The most common prefixes are:

mega-	(symbol M)	meaning million
kilo-	(symbol k)	meaning thousand
deci-	(symbol d)	meaning tenth
centi-	(symbol c)	meaning hundredth
milli-	(symbol m)	meaning thousandth

17

NOTE: Please note that these symbols, as well as those mentioned further on, have been accepted internationally and are therefore the same in all languages. For this reason it is absolutely essential that they be used correctly in order to avoid confusion, for example in the case of mega- (M) and milli- (m).

Length, Distance. The base unit of length is the *metre* (symbol m). The metre is somewhat longer than a yard, and is therefore equal to a long pace.

Most Important Units of Length

kilometre (km)	metre (m)	millimetre (mm)
1 000 m	1 m	$\frac{1}{1\,000}$ m

In addition the centimetre (1 cm $= \frac{1}{100}$ m) is also used for body measurements and in the textile and clothing industry.

Practical Application

kilometre: Replaces the mile.
Road maps will indicate distances in kilometres and eventually speedometers will also be graduated in kilometres.

metre: Replaces yard and foot.
The dimensions of a room will be expressed in metres, e.g. 4 m × 5 m.
The dimensions of swimming-pools, sports fields, and tennis courts will be in metres, as well as distances in athletics. The 800 m sprint replaces the 880 yd sprint. The metre will be used in plans for buildings as well as in other industries. The housewife will buy her dress material by the metre and decimal parts of a metre.

centimetre: Replaces foot and inch.
All body measurements, chest, waist and hip measurements, will be in centimetres. The well-known 36-24-36 will change to 92-61-92.
Your height will also be measured in centimetres. A lady of average height will now be 163 cm tall. Measure your own height in centimetres, and use it as a basis for evaluating other heights you come

18

across. The centimetre will also be used in the textile industry. The width of material, seam lengths in dress patterns, as well as knitting instructions, will be given in centimetres. If you therefore own a metric measuring tape, you will have no problems if you follow the metric instructions. You will even find it easier since you will no longer have to worry about quarters and eighths.

millimetre: Replaces the inch.
The millimetre will be used in engineering drawings and the lengths and diameters of screws, etc. will be given in millimetres. The gauge systems which are used at present to describe thicknesses of wire, sheet metal, etc. will all be replaced by the millimetre. In shops the diameters of pots, pans and baking-pans will be given in millimetres. The widths of ribbon, diameters of buttons and allowances for seams will also be expressed in millimetres.

Area. The main unit of area is the *square metre* (symbol m²). This represents the area of a square of which the dimensions are 1 m by 1 m.

Most Important Units of Area

hectare (ha)	square metre (m²)	square centimetre (cm²)
10 000 m²	1 m²	$\frac{1}{10\ 000}$ m²

NOTE: Hectare is a special term used for an area of 10 000 m² and may be thought of as representing a square of which each side measures 100   m.

Practical Application

hectare: Replaces acre.
The area of farms will be indicated in hectares.

square metre: Replaces square yard and square foot.
The floor area of a room or house will be given in square metres.

NOTE: Surveying and mapping, which includes the sizes of properties, property transactions and the registration of deeds will be in metric units. All new registrations of properties of which the original deeds of transfer are still in the old units, will be converted to metric when transfers take place.

square centimetre: Replaces square inch for small areas.

Volume, Capacity. The main unit is the *cubic metre* (symbol m³) and represents the volume of a cube of which each side is 1   m in length.

Most Important Units of Volume

cubic metre (m³)   cubic decimetre (dm³)   cubic centimetre (cm³)

$$1 \quad m^3 \qquad \frac{1}{1\ 000} \ m^3 \qquad \frac{1}{1\ 000\ 000} \ m^3$$

NOTE: A special name for the cubic decimetre is the litre (symbol *l*) and it is used to express the volume of liquids as well as the capacity of such appliances as refrigerators. The litre, too, has multiples and submultiples, e.g.:

megalitre (M*l*)   kilolitre (k*l*)   litre (*l*)   millilitre (m*l*)

1 000 000 *l*   1 000 *l*   1 *l* $\frac{1}{1\ 000}$ *l*

The following is derived from the above:

1   k*l* is the same as 1   m³

and 1   m*l* is the same as 1   cm³

Practical Application

cubic metre: Replaces cubic yard and cubic foot.

megalitre: Replaces gallon for large volumes of water.
               The water content of dams will be measured in megalitres.

kilolitre: Replaces gallon.
               Water used will be calculated by the amount of water, in kilolitres, consumed by the household.

litre: Replaces gallon and pint.
               You will come across the litre whenever you buy liquids commercially, 1 litre bottles of soft drinks, and milk will be sold in 1 litre containers.
               Gasoline for your car will be registered in litres on the pump.

millilitre:	Replaces the pint and fluid ounce.

The volume of liquids bought commercially in quantities of less than 1 litre, is given in millilitres. Apart from the above-mentioned 1 litre milk container, other liquids will be sold in 500 ml and 250 ml containers. The quantity of cooking oil, paint, motor oil, beer, spirits, soft drinks, etc. will be given in millilitres on the container.

Mass. The base unit of mass is the *kilogram* (symbol kg). The kilogram is about 10 per cent more than 2 lb. Half a kilogram of butter (500 grams) will therefore replace the existing pound of butter.

Most Important Units of Mass

megagram (Mg)	kilogram (kg)	gram(g)
1 000 000 g	1 000 g	1 g
1 000 kg	1 kg	$\frac{1}{1\,000}$ kg

NOTE: The megagram (Mg) is also known as a metric ton (symbol t). As soon as there can be no possible confusion, the metric ton will be known as ton only.

The difference between mass and weight. It is essential to distinguish between mass and weight. The *mass* of an object refers to the quantity of its matter or substance, and is expressed in kilograms. The *weight* of an object is the force with which it is attracted by the earth, and is expressed in newtons. The housewife is, therefore, interested in the mass of an object or commodity, e.g. the mass of butter she buys, while weight is mostly of interest to scientists and space travellers.

Practical Application

metric ton:	Replaces the existing ton for big loads.
kilogram:	Replaces 2 lb and over.

You will therefore buy rice, meat, sugar, fresh vegetables, etc. by the kilogram. Onions and potatoes probably in bags of 2, 5 and 10 kg respectively. Your body mass will be determined in kilograms.

gram: Replaces the pound and ounce.
                  Quantities less than 1   kg are given in grams. If
                  you need less than 1   kg of meat you will buy say
                  500   g. Spices will be sold in containers of 25 or
                  30   g. Butter will be marketed in quantities of 500
                  and 250   g. The mass of vegetables, fruit or jam
                  in tins will be given in grams. The gram will also be
                  used for knitting-wool.

Velocity, Speed. Velocity is expressed in *kilometres per hour*
(symbol km/h).

NOTE: The "h" in "km/h" refers to the Latin word "hora,"
meaning hour, and has been accepted internationally as the symbol for hour.

Practical Application

kilometres per
  hour: Replaces miles per hour.

                  The speedometers of all cars in the future will be
                  graduated in km/h. The speed limits may be as
                  follows

                  School zones   — 40   km/h
                  City & town   — 48   km/h
                  Caution areas   — 56   km/h
                  Regular   — 64   km/h
                  Secondary road   — 80   km/h
                  Highway   — 96   km/h

                  Do not be fooled by these speeds for they are
                  equal to only about 25, 30, 35, 40, 50, and 60
                  m.p.h. respectively.

Pressure. For all practical purposes, pressure will be expressed in
bars (symbol bar). The bar is very nearly equal to atmospheric
pressure at sea-level.

Most Important Units of Pressure

       bar (bar)   millibar (mbar)
         1   bar $\frac{1}{1\,000}$ bar

Practical Application

bar: Replaces pound-force per square inch.
The bar is used for pressure in pressure cookers as well as for tire pressures. The latter pressures will vary from 1,4 to 2 bars.

millibar: is an established unit already, and is mainly used in aeroplanes to determine altitude. You have most probably come across the millibar in weather reports in which barometric pressures are expressed in millibars.

Temperature. In practice temperature is expressed in *degrees celsius* (°C). The zero point on this scale is the temperature at which water freezes, while the boiling point of water is 100° C.

Practical Application

Celsius: Replaces Fahrenheit. Atmospheric and body temperatures are expressed in °C. Oven temperatures for baking purposes will also be given in °C.

Energy. The basic unit of energy is the *joule* (symbol J). As in the case of the calorie, the joule is too small for everyday use and the kilojoule will therefore be used.

Practical Application

kilojoule: Replaces the kilocalorie.
The kilojoule is used mainly in dietetics.

GENERAL REMARKS

Symbols. As has been mentioned, every metric unit has one, and only one, correct symbol. The symbols were chosen with great care on international level in order that no confusion about the symbols for units could ever arise. They remain the same in all languages in the world.

Symbols are never followed by a period
i.e. 15 kg and *not* 15 kg.

23

No addition such as "s" should be made to the symbols to indicate plurality,

i.e. 12 cm and *not* 12 cms

Remember: "s" is the symbol for second.

A space is always left between a figure and the subsequent symbol, eg. 15 km *and not* 15km.

In the spoken language, as well as when the unit is written out in full, the correct plural form in English is, for instance, "20 metres."

The use of the comma as a decimal sign in practice. We have always used the period as a decimal sign, however, the comma is used in many countries abroad with established metric systems. In all likelihood the comma may eventually supplant the period universally.

Sequences, co-ordinates and other sets. Since confusion may arise if the comma is used to separate elements of sequences, co-ordinates etc. if these elements already contain the comma as a decimal sign, it is recommended that the semicolon should be used consistently to separate elements. This is also in line with the usage in metric countries.

Examples

Sequences

(a) Decimals: 1,5; 2,0; 2,5; 3,0; . . .
(b) Whole numbers: 1; 2; 4; 8; 16; . . .

Co-ordinates

(a) Decimals: (1,75; 2,80)
(b) Whole numbers: (2; 4)

The method of writing units. The names of all units are in lower case letters when written out in full, e.g. kilometre, gram, litre, etc. The symbols for most units are also in lower case letters, e.g. km, g, *l*, etc. Only in cases where the name of a unit is derived from the name of a person, should the symbol be either a capital letter or commence with a capital letter if the symbol consists of more than one letter, e.g.

	gram,	symbol g
but	joule,	symbol J

APPROXIMATE CONVERSION FACTORS

The conversion factors below are approximate, and should not be used where a high degree of accuracy is required.

Try to use the conversion factors as seldom as possible, rather make an attempt to think metric. Should you, however, need a conversion factor, e.g. for an imperial recipe, the conversion factor should be used once only. The metric quantity can then be written next to the imperial quantity and after that should be used always. A continual conversion of units will complicate the change-over considerably.

Length, Distance

	Metric to non-metric		Non-metric to metric
	1 km $=$ 0,6 mile		1 mile $=$ 1,6 km
or	8 km $=$ 5 miles	or	5 miles $=$ 8 km
	1 m $=$ 1,1 yd		1 yd $=$ 0,9 m
	or 3,3 ft		1 ft $=$ 0,3 m
	or 39,4 in		1 inch $=$ 25,4 mm
	1 mm $=$ 0,04 inch		
	$(= \frac{1}{25}$ inch$)$		

Area

	Metric to non-metric	Non-metric to metric
1 ha $=$ 2,5 acres	1 acre $=$ 0,4 ha
1 m² $=$ 1,2 sq yd	1 sq yd $=$ 0,8 m²
$=$ 10,8 sq ft	1 sq in $=$ 6,5 cm²
1 cm² $=$ 0,2 sq in	(or $6\frac{1}{2}$ cm²)

Volume, Capacity

Metric to non-metric	Non-metric to metric
1 m³ = 1,3 cub yd 　　　 or 35,3 cub ft	1 cub yd = 0,8 m³ 1 cub inch = 16,4 cm³
1 cm³ = 0,06 cub inch	1 gallon = 4,5 l
1 l = 0,2 gallon	(or 4$\frac{1}{2}$ l)
= 1,75 pt	1 pint = 567 ml
(or 1$\frac{3}{4}$ pt)	or 7 pt
or 4 l = 7 pt	= 4 l
1 ml = 0,04 fluid oz	1 fluid oz = 28,4 ml

Mass

Metric to non-metric	Non-metric to metric
1 metric ton = 2 205 lb	1 ton = 0,9 metric ton
= 1,1 tons	1 lb = 0,45 kg
1 kg = 2,2 lb	= 454 g
(or	1 oz = 28,3 g
2$\frac{1}{5}$ lb)	
1 g = 0,04 oz	

Velocity, Speed

Metric to non-metric	Non-metric to metric
1 km/h = 0,6 m.p.h.	1 m.p.h. = 1,6 km/h
or 8 km/h = 5 m.p.h.	or 5 m.p.h. = 8 km/h

Pressure

Metric to non-metric	Non-metric to metric
1 bar = 14,5 pound- 　　　 force per 　　　 sq inch	1 pound- force per sq inch = 0,07 bar

Temperature

$10° C = 50° F$. For every increase or decrease of 5 degrees on the Celsius scale, 9 degrees are added or subtracted from the Fahrenheit scale.

or Number of degrees Celsius
$$= (\text{Number of degrees Farenheit} - 32) \times \tfrac{5}{9}$$
Number of degrees Fahrenheit
$$= (\text{Number of degrees Celsius} \times \tfrac{9}{5}) + 32$$

Energy

Remember: The joule (J) replaces the calorie (cal) and the kilojoule (kJ) replaces the kilocalorie (kcal)

Metric to non-metric	Non-metric to metric
1 J = 0,24 cal	1 cal = 4,2 J
1 kJ = 0,24 kcal	1 kcal = 4,2 kJ

Force, Weight

Metric to non-metric	Non-metric to metric
1 N = 0,225 pound-force (newton)	1 pound-force = 4,45 N

Work

Metric to non-metric	Non-metric to metric
1 kW = 1,34 horse-power (kilowatt)	1 horse-power = 0,75 kW

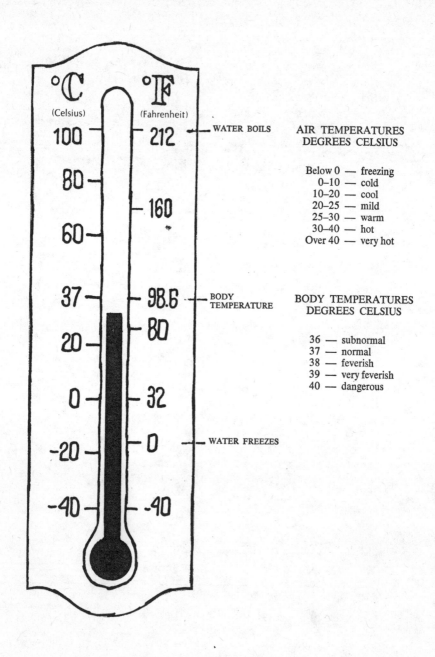

°C (Celsius)

100
80
60

37
20

0

-20

-40

°F (Fahrenheit)

212 — WATER BOILS

160

98.6 — BODY TEMPERATURE
80

32

0 — WATER FREEZES

-40

AIR TEMPERATURES
DEGREES CELSIUS

Below 0 — freezing
0–10 — cold
10–20 — cool
20–25 — mild
25–30 — warm
30–40 — hot
Over 40 — very hot

BODY TEMPERATURES
DEGREES CELSIUS

36 — subnormal
37 — normal
38 — feverish
39 — very feverish
40 — dangerous

IV. THE EASY WAY TO LEARN THE METRIC SYSTEM

There follow some easy step-by-step exercises that will lead to a better understanding of the metric system. If you can add, subtract, and divide in decimals you are well on your way. The following tables of approximate equivalents are of great practical use to teachers at all levels. They place at their disposal, at once, sufficient material dealing with metric measurements of length, area, volume, and mass to enable them to devise any number of drill exercises, work assignments, diagnostic tests, and periodic proficiency examinations to evaluate the progress being made by any group or individual in developing metric skills thereby discerning the weaker areas to be concentrated on.

In the ensuing pages the busy teacher is given a considerable amount of material in condensed form including some easy practice examples with answers at the end of the book. These may be used as patterns for developing many more. The teacher is given a considerable amount of valuable data at least cost in time and energy eliminating many hours of tedious reading, preparation of classroom work, assignments and correction of papers, and other time-consuming routine work. The condensed tables, practice examples, and other material which follows are very simple and they are ready for prompt use with little prior preparation required on the part of the teacher. They serve as an unlimited source of reference and drill material with the answers contained in the text itself.

Approximate Conversion Factors for Making Quick Calculations

To convert	to	Multiply by
LENGTH		
inches	millimetres	25
inches	centimetres	2.5
feet	centimetres	30
feet	metres	0.3
yards	centimetres	90
yards	metres	0.9
rods	metres	5
miles	kilometres	1.6
millimetres	inches	0.04
centimetres	inches	0.4
centimetres	feet	0.033
metres	inches	40
metres	feet	3.3
metres	yards	1.1
metres	rods	0.2
kilometres	miles	0.62
AREA		
square inches	square centimetres	6.5
square feet	square metres	0.09
square yards	square metres	0.84
acres	hectares	0.4
square miles	square kilometres	2.6
square centimetres	square inches	0.16
square metres	square feet	11
square metres	square yards	1.2
hectares	acres	2.5
square kilometres	square miles	0.4
VOLUME		
cubic inches	cubic centimetres	16.4
cubic feet	cubic metres	0.03
cubic yards	cubic metres	0.76
cubic centimetres	cubic inches	0.06
cubic metres	cubic feet	35
cubic metres	cubic yards	1.3

CAPACITY (dry measure)

bushels	litres	35
litres	bushels	0.03

CAPACITY (liquid)

fluid ounce	millilitres	30
pints	millilitres	473
pints	litres	0.47
quarts	litres	0.95
gallons	litres	3.8
millilitres	fluid ounces	0.034
millilitres	pints	0.002
litres	pints	2.1
litres	quarts	1.06
litres	gallons	0.26

MASS

ounces	grams	28
ounces	kilograms	0.028
pounds	grams	454
pounds	kilograms	0.45
tons	metric tons	0.9
grams	ounces	0.035
grams	pounds	0.002
kilograms	ounces	35
kilograms	pounds	2.2
metric tons	tons	1.1

TEMPERATURE

degrees Fahrenheit	degrees Celsius	$\frac{5}{9}$ after subtracting 32
degrees Celsius	degrees Fahrenheit	$\frac{9}{5}$ then add 32

Examples:

$$\text{(Fahrenheit degrees} - 32 \times \tfrac{5}{9} = \text{Celsius degrees)}$$
$$\text{(Celsius degrees} \times \tfrac{9}{5} + 32 = \text{Fahrenheit degrees)}$$

To convert
To convert
To convert

to
to
to

Multiply by
Multiply by
Multiply by

FUNDAMENTAL UNITS, PREFIXES, VALUES AND SYMBOLS

Fundamental Units: Length: the metre (m)
Mass: the gram (g)
Volume: the litre (*l*)

Prefixes used with fundamental units

MULTIPLES (larger than base units)	Numerical Value	Symbol used when adding prefixes to:		
		metre (m)	litre (*l*)	gram (g)
Base units				
mega- (M) (one million times)	1,000,000	Mm	M*l*	Mg
kilo- (k) (one thousand times)	1,000	km	k*l*	kg
hecto- (h) (one hundred times)	100	hm	h*l*	hg

SUBMULTIPLES (smaller than base units)				
deci- (d) one-tenth	0.1	dm	d*l*	dg
centi- (c) one-hundredth	0.01	cm	c*l*	cg
milli- (m) one-thousandth	0.001	mm	m*l*	mg
micro- (μ) one-millionth	0.000 001	μm	μ*l*	μg
nano- (n) one-billionth	0.000 000 001	nm	n*l*	ng

Using the prefixes: Here is how to use the prefixes with the base units:

Weight: kilogram (kg)
             = one thousand grams (1,000   g)
Length: millimetre (mm)
             = one-thousandth of a metre (1/1,000   m)
Volume: centilitre (c*l*)
             = one-hundredth of a litre (1/100   *l*)
             decilitre (d*l*)
             = one-tenth of a litre (1/10   *l*)

Measurement of Length and Distance

The base unit of length is the metre. Lengths greater than one metre are called multiples and they are expressed as follows:

1 kilometre (km)	=	1,000 metres
1 hectometre (hm)	=	100 metres
1 dekametre (dkm)	=	10 metres

Lengths smaller than one metre (fractions) are called submultiples and they are expressed as follows:

$$1 \text{ decimetre (dm)} = 1/10 \text{ or } 0.1 \text{ metre}$$
$$1 \text{ centimetre (cm)} = 1/100 \text{ or } 0.01 \text{ metre}$$
$$1 \text{ millimetre (mm)} = 1/1000 \text{ or } 0.001 \text{ metre}$$

Practice examples:

1. a. How many millimetres make
      one centimetre? _____ mm
      one decimetre? _____ mm
      one metre? _____ mm
   b. How many centimetres make
      one decimetre? _____ cm
      one metre? _____ cm
   c. How many decimetres make
      one metre? _____ dm
      0.75 metre? _____ dm
      10.5 metres? _____ dm

2. a. 35 dm = _____ m?   f. 375 cm = _____ m?
   b. 8 cm = _____ mm?   g. 37.5 dm = _____ m?
   c. 3,250 mm = _____ m?   h. 1.75 m = _____ mm?
   d. 7.5 m = _____ mm?   i. 3,250 mm = _____ m?
   e. 8.25 cm = _____ mm?   j. 200 dm = _____ m?

3. Write the following in decimal figures.

   a. 1 metre, 7 decimetres, 5 centimetres, 15 millimetres.
      Answer: _____
   b. 8 decimetres, 12 centimetres, 5 millimetres.
      Answer: _____
   c. 12 metres, 12 millimetres.
      Answer: _____

4. a. How many millimetres are there in 9.7 metres?

      Answer: _____ mm
   b. How many centimetres are there in 7.275 metres?
      Answer: _____ cm
   c. How many decimetres are there in 33.375 metres?
      Answer: _____ dm

5. Measure the width of your desk and give it as follows:

   a. in millimetres ————
   b. in centimetres ————
   c. in decimetres ————
   d. in metres ————

6. Measure the height of the door to this room and give it as follows:

   a. in metres, centimetres, and millimetres ————
   b. in decimal figures ————

7. Convert the following:

   a. 350 mm to cm ————
   b. 50 dm to mm ————
   c. 10 cm to mm ————
   d. 1000 mm to dm ————
   e. 250 cm to dm ————
   f. 500 mm to cm ————

8. Write the following in decimal figures.

   a. 6 dkm, 8 cm = ————
   b. 9 m, 7 dm, and 9 cm = ————
   c. 7 m, 5 dm, 4 cm = ————
   d. 18 km = ————

9. Measure the length and width of the room and give its area in square metres (m^2). (length $\times$ width = area.)
Answer: ————

10. Write in decimals and add 3 m, 5 dm, 30 cm, 750 mm.
Answer: ———— m total.

Measurement of Volume:

The base unit for volume is the litre. Quantities greater than one litre are called multiples and they are expressed as follows:

1 kilolitre (k*l*)	=	1000 litres
1 hectolitre (h*l*)	=	100 litres
1 dekalitre (dk*l*)	=	10 litres

Quantities smaller than one litre (fractions) are called submultiples and they are expressed as follows:

35

$$
\begin{array}{lll}
1 \text{ decilitre (d}l) & = & 1/10 \text{ or } 0.1 \text{ litre} \\
1 \text{ centilitre (c}l) & = & 1/100 \text{ or } 0.01 \text{ litre} \\
1 \text{ millilitre (m}l) & = & 1/1000 \text{ or } 0.00 \text{ litre}
\end{array}
$$

thus,

$$
\begin{array}{lll}
10 \text{ decilitres} & = & 1 \text{ litre} \\
100 \text{ centilitres} & = & 1 \text{ litre} \\
1000 \text{ millilitres} & = & 1 \text{ litre}
\end{array}
$$

Give the following answers:

How many millilitres make 1 centilitre? _____

How many centilitres make 1 decilitre? _____

How many decilitres make 1 litre? _____

Examples:

a. If one thousand grains of oats weigh 1 gram (g), then each grain weighs 1 milligram (mg) or 1/1,000th of a gram.

b. If one hundred grains of corn weigh 1 gram (g), then each grain weighs 1 centigram (cg) or 1/100th of a gram.

c. If ten pebbles weigh one gram, then each pebble weighs 1 decigram (dg) or 1/10th of a gram.

Practice examples:

1. a. How many milligrams make 1 gram? _____
   b. How many centigrams make 1 gram? _____
   c. How many decigrams make 1 gram? _____
   d. How many grams make 1 kilogram? _____

2. a. 500 mg = _____ cg
   b. 225 cg = _____ dg
   c. 65 dg = _____ g
   d. 3 g   = _____ mg
   e. 4 g   = _____ cg
   f. 5 g   = _____ dg

3. a. 800 ml = _____ cl
   b. 7 dl = _____ cl
   c. 5 cl = _____ ml
   d. 9 dl = _____ cl
   e. 3.9 dl = _____ ml
   f. 3.7 cl = _____ ml

Measurement of Mass Quantity

The base unit for mass quantity is the gram (g). Multiples of the gram are expressed as follows:

1 kilogram (kg)	=	1000 grams
1 hectogram (hg)	=	100 grams
1 dekagram (dkg)	=	10 grams

Submultiples (fractions) of grams are expressed as follows:

1 decigram (dg)	=	1/10 or 0.1 of a gram
1 centigram (cg)	=	1/100 or 0.01 of a gram
1 milligram (mg)	=	1/1000 or 0.001 of a gram

thus,

10 decigrams	=	1 gram (g)
100 centigrams	=	1 gram
1000 milligrams	=	1 gram

Stated inversely, mass quantities of more than 1 gram have the following names:

10 grams	=	1 dekagram (dkg)
100 grams	=	1 hectogram (hg)
1000 grams	=	1 kilogram (kg)

Mass quantities of less than one gram have the following names:

1/10 of a gram	=	1 decigram (dg)
1/100 of a gram	=	1 centigram (cg)
1/1000 of a gram	=	1 milligram (mg)

Approximate equivalents of miles and kilometres (km)

Miles	1	2	3	4	5	6	7	8	9	10
Kilometres	1.6	3.2	4.8	6.4	8.0	9.6	11.2	12.8	14.4	16.0

Length

Take any automobile road map, look at the mileage chart and change the distances from miles to kilometres. The mile is actually equal to 1.6093 kilometres but for practical purposes we will use the above approximate equivalents.

Fact: 1 mile =   1.6 km

Method: miles × 1.6 gives kilometres (km)

Practice examples:

1. Akron to Atlanta   668 miles $\times$ 1.6   = _____ km
2. Boston to Birmingham   1190 miles
   $\times$ 1.6   = _____ km
3. Chicago to Jacksonville   1006 miles
   $\times$ 1.6   = _____ km
4. Detroit to Tampa   1174 miles $\times$ 1.6   = _____ km
5. Indianapolis to Montreal   862 miles
   $\times$ 1.6   = _____ km
6. Memphis to Boston   1326 miles
   $\times$ 1.6   = _____ km
7. New York to Louisville   758 miles
   $\times$ 1.6   = _____ km
8. Washington, D.C. to Miami   1105
   miles $\times$ 1.6   = _____ km
9. Chicago to Raleigh   790 miles
   $\times$ 1.6   = _____ km
10. Milwaukee to Philadelphia   828 miles
    $\times$ 1.6   = _____ km

More practice examples can be worked up from the above tables, e.g. A car rental agency charges 12¢ per mile, how much is that per kilometre? How much will the charge be from _____ to _____ ?

NOTE: See the index for tables of accurate equivalents.

Approximate equivalents yards and metres (m)

Yards	1	2	3	4	5	6	7	8	9
Metres	0.9	1.8	2.7	3.6	4.5	5.4	6.3	7.2	8.1

Yards to Metres (m)

The yard is actually equal to 0.9144 metre but for practical purposes we will use the above approximate equivalents.

Fact: 1 yard = 0.9 m

Method: yards $\times$ 0.9 gives metres (m)

Practice examples:

1. The 100 yard dash is how many metres?
   100 × 0.9   = _____ m
2. A kite string is let out 145 yards. How many metres is that?   145 × 0.9   = _____ m
3. A swimming pool is 8 yds. × 12 yds. What is its perimeter in metres?
   40 × 0.9   = _____ m
4. He ran 82 yds. for a touch-down. How many metres did he run?   82 × 0.9   = _____ m
5. A driveway is $25\frac{1}{2}$ yds. long. How many metres is that?   25.5 × 0.9   = _____ m
6. A piece of goods is $1\frac{1}{3}$ yds. wide. What is that in metres?   1.333 × 0.9   = _____ m
7. Our lawn is 20 yds. wide by 40 yds. long. Give that in metres.   20 × 0.9   = _____ m
   and 40 × 0.9   = _____ m
8. The 1972 Olympic shot-put record was $23\frac{1}{6}$ yds. Give that in metres.
   23.16 × 0.9   = _____ m
9. The discus throw went $72\frac{2}{3}$ yds. How many metres is that?   72.66 × 0.9   = _____ m
10. The shelf is $2\frac{1}{2}$ yds. from the floor. What is that in metres?   2.5 × 0.9   = _____ m

NOTE: See the index for tables of accurate equivalents.

Approximate equivalents feet and centimetres (cm)

Feet	1	2	3	4	5	6	7	8	9
Centimetres	30	61	91	122	152	183	213	244	274

Feet to Centimetres (cm)

The foot is actually equal to 30.48 centimetres but for practical purposes we will use the above approximate equivalents.

Fact: 1 foot   =   30   cm

Method: feet   ×   30 gives centimetres (cm)

39

Practice examples:

1. He is 6 ft. 6 in. tall. How many centimetres is that?   6.5 × 30   = _____ cm

2. The distance from 1st to 2nd base is
   30 ft. How many centimetres is that?
   30 × 30   = _____ cm

3. A boxer skips a 12-foot rope. Give
   that in centimetres.   12 × 30   = _____ cm

4. A desk is 24 in. × 60 in. What is its
   perimeter in centimetres?
   2 + 5 + 2 + 5 = 14 × 30   = _____ cm

5. The pitcher's box is 90 feet from home
   plate. What is that in centimetres?
   90 × 30   = _____ cm

6. The aisle is 2½ ft. wide. How many
   centimetres is that?   2.5 × 30   = _____ cm

7. How many centimetres is a pen 6
   inches long?   0.5 × 30   = _____ cm

8. A window is 3 ft. 6 in. wide by 4 ft.
   4 in. high. Give these dimensions in   = _____ cm
   centimetres.   3.5 × 30; 4.33 × 30   = _____ cm

9. A swimming pool is 9 ft. at the deep
   end. How many centimetres is that?
   9 × 30   = _____ cm

10. The golfer missed a 9-inch putt. What
    is that in centimetres?   = _____ cm

NOTE: See the index for tables of accurate equivalents.

Approximate equivalents inches to centimetres (cm)

Inches	1	2	3	4	5	6	7	8	9
Centimetres	2.5	5.0	7.6	10.0	12.7	15.0	17.8	20.0	22.9

Inches to Centimetres (cm)

One inch is actually equal to 2.54 centimetres but for practical
purposes we will use the above approximate equivalents.

Fact: 1 inch = 2.5 cm

Method: inches × 2.5 gives centimetres (cm)

Practice examples:

1. Give the length of your pencil in centimetres. _____ cm
2. Give your chest measurement in centimetres. _____ cm
3. Give the length of your belt in centimetres. _____ cm
4. Measure the width of your desk and give it in centimetres. _____ cm
5. Measure your shoe with a ruler and state how many centimetres long it is. _____ cm
6. The ruler broke in two pieces, one $3\frac{1}{4}$ inches long, the other $8\frac{3}{4}$ inches long. Give the length of both pieces in centimetres. _____ cm _____ cm
7. Give the measurement of your hand-span in centimetres. _____ cm
8. How many centimetres long is a fishing rod measuring $6\frac{1}{2}$ ft.? _195_____ cm
9. The tackle box is 12 in. long, 8 in. wide, and 6 in. high. Give these measurements in centimetres. _____ cm _____ cm _____ cm
10. A wrestler's neck measures $20\frac{1}{2}$ inches in circumference. What is that in centimetres? _____ cm

NOTE: See the index for tables of accurate equivalents.

Approximate equivalent of inches to millimetres (mm)

Inches	1	2	3	4	5	6	7	8	9
Millimetres	25	50.8	76.2	101.6	127	152.4	177.8	203.2	228.6

Inches to Millimetres (mm)

Millimetres are for measuring smaller things. Thus, let us now change inches to millimetres. One inch is actually equal to 25.4 millimetres but for practical purposes we will use the approximate equivalents given above.

Fact: 1 inch   =   25   mm

Method: inches ×   25 gives millimetres (mm)

Practice examples

1. A piece of paper is 9 in. $\times$ 12 in.
   State these measurements in milli-   = _____ mm
   metres.   9 $\times$ 25; 12 $\times$ 25   = _____ mm
2. Measure your pencil and give it in
   millimetres.   = _____ mm
3. Peter sharpened his pencil down to
   $4\frac{3}{4}$ inches. How many millimetres is
   that?   4.75 $\times$ 25   = _____ mm
4. Janie is $1\frac{1}{2}$ inches taller than Liza.
   How many millimetres taller is she?   = _____ mm
5. The thickness of a board is $\frac{1}{2}$ inch.
   State that in millimetres.   0.5 $\times$ 25   = _____ mm
6. Measure the width of your shoe and
   give it in millimetres.   = _____ mm
7. What is the circumference of your
   neck in millimetres?   = _____ mm
8. A pencil box is 3 inches wide by 8
   inches long. What is its perimeter in
   millimetres?   3 + 8 + 3 + 8
   = 22 $\times$ 25   = _____ mm
9. How wide is a $1\frac{1}{2}$-inch belt in millimetres?   1.5 $\times$ 25   = _____ mm
10. A line $7\frac{1}{4}$ inches long is how many
    millimetres?   7.25 $\times$ 25   = _____ mm

NOTE: See the index for tables of accurate equivalents.

Practice examples:

A dressmaker has received the following orders for garments.
Change the measurements to metric equivalents.

Standard Measurements

Order No. 1—Size 12

Bust 34 inches = _____ cm

Waist $25\frac{1}{2}$ inches = _____ cm

Hip 36 inches = _____ cm

Back waist length

$16\frac{1}{2}$ inches = _____ cm

Goods required

$2\frac{7}{8}$ yds. = _____ m

Order No. 2—Size 14

Bust 36 inches = _____ cm

Waist 27 inches = _____ cm

Hip 38 inches = _____ cm

Back waist length

$16\frac{1}{2}$ inches = _____ cm

Goods required

$2\frac{7}{8}$ yds. = _____ m

Order No. 3—Size 16

Bust 38 inches = _950_ cm
Waist 29 inches = _____ cm
Hip 40 inches = _____ cm
Back waist length
$16\frac{3}{4}$ inches = _____ cm
Goods required
$2\frac{7}{8}$ yds. = _____ m

Order No. 4—Size 18

Bust 40 inches = _____ cm
Waist 31 inches = _____ cm
Hip 42 inches = _____ cm
Back waist length
17 inches = _____ cm
Goods required
$2\frac{7}{8}$ yds. = _____ m

Order No. 5—Size 40

Bust 44 inches = _____ cm
Waist 36 inches = _____ cm
Hip 46 inches = _____ cm
Back waist length
$17\frac{3}{8}$ inches = _____ cm
Goods required
3 yds. = _____ m

Order No. 6—Size 42

Bust 46 inches = _____ cm
Waist 38 inches = _____ cm
Hip 48 inches = _____ cm
Back waist length
$17\frac{1}{2}$ inches = _____ cm
Goods required
$3\frac{1}{4}$ yds. = _____ m

AREA MEASUREMENTS

Fact: length $\times$ width = area

When measurements are given in inches, feet, yards, or miles, to obtain square centimetres, square metres, or square kilometres we must first change them to metric equivalents and then multiply.

Method: Again using approximate factors,

1 inch	=	2.5	cm	therefore inches	$\times$	2.5	=	cm
1 foot	=	30	cm	therefore feet	$\times$	30	=	cm
1 foot	=	0.3	m	therefore feet	$\times$	0.3	=	m
1 yard	=	0.9	m	therefore yards	$\times$	0.9	=	m
1 mile	=	1.6	km	therefore miles	$\times$	1.6	=	km

Examples:

a. A rectangle 10 inches $\times$ 12 inches is the same as 25 cm $\times$ 30 cm, i.e. (10 $\times$ 2.5 and 12 $\times$ 2.5) 25 cm $\times$ 30 cm = 750 cm² area.

43

b. A rectangle 10 feet × 20 feet is the same as 3 m × 6 m, i.e. (10 × 0.30 and 20 × 0.30) 3 m × 6 m = 18 m² area.

c. A rectangle 10 yards × 15 yards is the same as 9 m × 13.5 m, i.e. (10 × 0.9 and 15 × 0.9) 9 m × 13.5 m = 121.5 m² area.

d. An area 10 miles wide by 15 miles long is the same as 16 km × 24 km, i.e. (10 × 1.6 and 15 × 1.6) 16 km × 24 km = 384 km² area.

NOTE: See the index for complete tables of area measurements.

The gourmet shelf in the delicatessen has the following imported items packed in grams. Give their American equivalents. (Use preceding table of approximate equivalents.)

1. Chili powder   56.70 grams = _____ oz.
2. Paprika   42.25 grams = _____ oz.
3. Ground cinnamon   35.44 grams = _____ oz.
4. Mace   31.89 grams = _____ oz.
5. Nutmeg   63.79 grams = _____ oz.
6. Caviar   454 grams   = _____ oz.
7. Biscuits   1,350 grams = _____ oz.
8. Rusk   269 grams   = _____ oz.
9. Shortbread   212.6 grams = _____ oz.
10. Ceylon Tea   227 grams   = _____ oz.

The manager of the super-market ordered the following items from the warehouse. Give their metric equivalents.

11. Granulated sugar   440 lbs.   = _____ kg
12. Potatoes   385 lbs.   = _____ kg
13. Onions   330 lbs.   = _____ kg
14. Apples   661 lbs.   = _____ kg
15. Flour   1,000 lbs. = _____ kg

MISCELLANEOUS PRACTICE EXAMPLES

Convert the following:

1. 1 loaf of bread, 16 ounces to _____ grams (g)
2. 2 lbs. butter   to _____ grams (g)
3. 5 lbs. sugar   to _____ kilograms (kg)

4. 10 lbs. potatoes   to _____ kilograms (kg)
5. 1 hand of bananas, $2\frac{3}{4}$ lbs. to _____ kilograms (kg)
6. $3\frac{1}{2}$ lbs. spinach   to _____ kilograms (kg)
7. 1 cannister black pepper,
    4 oz.   to _____ grams (g)
8. 1 can frozen strawberries,
    8 oz.   to _____ grams (g)
9. $\frac{1}{2}$ gallon of milk   to _____ litres (*l*)
10. 1 quart jar of honey   to _____ litres (*l*)
11. $1\frac{1}{2}$ quart of apple juice   to _____ litres (*l*)
12. 1 gallon of cider   to _____ litres (*l*)
13. 5 yards of cheese cloth   to _____ metres (m)
14. $3\frac{1}{2}$ yards of muslin   to _____ metres (m)
15. 15 yards curtain material   to _____ metres (m)

THE CHEF'S PROBLEMS

Practice examples:

1. A ham weighing 5 kg shrank 10% in cooking.
How many grams did it shrink? _____ g
2. A meat loaf weighing 1 kg lost 5% in cooking.
How many grams did it lose? _____ g
3. The chef ordered the following:
5 kg bacon
10 kg ham
22.5 kg ground beef
500 grams baking powder
25 kg cake flour
250 g powdered cinnamon
125 g powdered nutmeg
10.5 kg butter
5 kg salt
What is the total weight of the order? _____ kg
4. The butcher dressed a hind of beef weighing 250 kg. He
took off 15.5 kg of fat, and 33.75 kg of bones.
What was the weight of the meat remaining? _____ kg
5. The chef ordered 25 kg of assorted fish. He found the de-
livery was short 2.300 kg of flounder, 535 g short of scal-
lops, and 750 g short of mackerel.
What was the total shortage? _____ kg

What was the total weight of fish he actually got?
_____ kg

6. Eight litres of water are needed to cook 1 kilogram of macaroni. After cooking there remained only 5 litres.
How many millilitres of water had evaporated?
_____ ml
How many centilitres? _____ cl

7. The baker mixed a batch of bread dough weighing 5 kilograms. If he made loaves weighing 500 grams each, how many loaves would he get? _____ loaves.

8. 5 kilograms of potatoes lost 1 kg 450 g in peeling them. What was the remaining weights? _____ kg

9. A loaf of French bread is 0.75 metre long.
If it is cut into 5 equal pieces how many centimetres would there be in each piece? _____ cm

10. If you drove your car in Europe how would you read the following speeds in U.S. miles?
65 km/h; 80 km/h; 96 km/h?
_____ mph _____ mph _____ mph

11. Give today's temperature in degrees Celsius. _____ °C

12. A recipe for baking a cake calls for a 350 degree Fahrenheit oven heat. What is that in degrees Celsius? _____ °C

V. GLOSSARY OF METRIC TERMINOLOGY AND REVISIONS IN METRIC TECHNICAL LANGUAGE

This glossary contains SI units, symbols, and prefixes as well as other units used in conjunction with the SI, and records their notation in English. These symbols, which are based on those adopted by the Conférence Générale des Poids et Mesures (CGPM) and the International Standards Organization (ISO), will be the only correct ones.

The glossary does not include all possible compounds; it contains representative examples only, which will enable the reader to make his own compounds. It also contains some units, such as "minute" and "kilowatt hour," which are not included in the SI, but which, for practical reasons still prevail.

According to international usage, the symbol of a unit is not followed by a full stop, e.g. m (metre), kg (kilogram) and l (litre). Although some of the symbols are written with capital letters, the units when written in full should be in lower case letters, e.g. J = joule, N.m = newton metre.

In this glossary the point is used to indicate multiplication, e.g. m.s, and the solidus to indicate division, e.g. m/s^2. The function of the negative exponent should also be noted, e.g. $m.s^{-2}$ represents metre per second squared.

In terms of ISO Recommendation R31, a space instead of a point may, in the case of symbols, be used to indicate multiplication. The use of either a point or a space is therefore permissible, e.g. metre second may be indicated by m.s or m s (whereas ms represents millisecond).

Occasionally it has been necessary to include symbols, e.g. cm, which do not accord with the preference rules of the SI.

In English the "-s" must be used to signify the plural form in all terms written out in full (e.g. amperes per metre). The symbols however have no plural form.

METRIC TERMINOLOGY

1. The Difference Between Mass and Weight

Mass is a fundamental property of an object which, in popular language, can be described as the quantity of matter it contains. The weight of an object is the force by which the earth or any other heavenly body attracts it.

In the space-age we live in, it becomes clear that mass and weight are two entirely different concepts. Today it is common knowledge that astronauts on a voyage to the moon become weightless and that on the moon their weight is less than one fifth of their weight on earth. Their mass, however, remains unchanged everywhere.

2. Confusion in the Past

In the Imperial system the unit of mass was the pound, while the unit of weight was the pound-force. In everyday use the word "force" disappeared from the compound "pound-force," with the result that the word "pound" was used for the expression of both weight and mass. As a result of this, the public came under the impression that mass and weight were the same concept.

3. Mass and Weight in the Metric System

Any good measuring system must adhere among others to the following two rules:

a) Two or more basic units are not used to measure the same concept.

b) A single unit is not used to measure two different concepts.

It is clear that when rule (a) is ignored, an unnecessary variety of units is used, and that infringement of rule (b) inevitably leads to confusion.

The modern metric system, the Système International d'Unités (SI), which has been adopted respects the above mentioned principles. It is for this reason that it is such an efficient and simple system.

The SI rightly distinguishes between the units of mass and weight. The kilogram (kg) is defined as the mass of the international prototype kilogram, in the safekeeping of the Bureau International des Poids et Mesures at Sèvres, France. Weight, being a force, is expressed in newtons (N) like all other forces.

By accepting the SI as its system of units, the United States simultaneously accepts the internationally recognized definitions of the kilogram (kg) and the newton (N).

4. Mass is Constant, Weight Varies

The weight of an object varies not only in space but also from one place to another on earth. If SI units are used, this fact can be expressed in figures. An object with a mass of 1   kg for instance has a weight of 9,809 26   N at Sèvres in France, or 9,806 65   N at sea level and 45 degrees northern latitude. The weight of the same object, however, is about 9,83   N at the poles and 9,77   N at the equator. These differences cannot be ignored.

The weight of a body varies with latitude and altitude, its mass does not.

5. Determining Mass

In daily practice, two methods are used to determine mass. For the first a balance is used. The object of which the mass has to be determined is placed on one pan of the balance while objects with a known mass are placed on the other pan until the scale is in balance, i.e. until the gravitational force exercised on the object of unknown mass (in other words its weight) is equal to the weight of the objects of known mass. If the weights are equal, the masses are equal. It is important to remember that, although gravitational force is used to compare the masses, *it is not necessary to know the value of the gravitational force to determine the mass.* The same method can be used on the moon, and it will give the same result, although the weights of objects there are entirely different from their weights on earth. The balance is therefore an instrument used to determine mass by using the force of gravity, i.e. weight, without knowing the value of the force of gravity.

Someone might ask: "What about the spring-balance? Surely the spring-balance measures the earth's gravitational force on an object, i.e. its weight." That is correct. In practice, however, spring-balances are calibrated in mass units almost without exception. When a sensitive spring-balance is taken from one place to another, before being used again, it is reassized by placing objects of known mass on it and adjusting it to give a correct reading of the mass. Therefore the assized spring-balance, too, is an instrument for determining mass and not weight.

All ordinary "weighing instruments" determine mass and not weight. A special experiment, or knowledge of the gravitational acceleration at a given place, is necessary to determine the weight of an object at such a place. A "weighing instrument" is thus actually a massmeter.

6. Masspieces

The standard objects used with a balance, up till now known as weights, have the characteristic that their masses are accurately known and clearly stamped on them, while their weight is unknown. They are used to determine the mass of other objects. It is therefore logical to speak of these objects as "masspieces" and not "weights." This term in addition enables one to distinguish clearly between the concept "mass" and the physical object which has a certain mass.

7. Application of the Concepts Mass and Weight

From the above it should be clear that in commercial transactions we are interested in mass and not in weight. It is the quantity of the matter that is being sold, and not the force by which the earth attracts the object or commodity concerned that interests both the buyer and the seller. The housewife wants to know the quantity of the butter she buys and not the force with which the earth attracts it. The physician is interested in the body mass of his patient and not in his weight, for when an astronaut becomes weightless in space, his doctor is not worried, but when his mass is reduced by a few kilograms, it may reflect a change in his state of health. In fact, it is mainly engineers and scientists who are interested in the term "weight" in its technical sense.

8. Use of Words

In the future we will speak of a "masspiece" instead of a "weight." A "weightpiece" is an object that may be used for testing purposes and on which the weight in newtons, as well as the place where it has that particular weight, is stamped.

We "measure" mass, like all other quantities such as length, area, volume, speed, voltage or temperature. To say that an object weighs 5 kg or that it has a weight of 5 kg makes no sense for the kilogram is not a unit of weight. It is correct to say: "the mass of the object is 5 kg." Expressions like "net mass 1 kg," "sold per mass unit" replaces "net weight 1 kg" and sold per weight unit" respectively. We cannot say: "I am 2 kg heavier." The correct expression is: "My mass has increased by 2 kg." In this connection it is only necessary to recall that the kilogram (kg) is used in expressing the concept "how much" in the sense of how much matter, whereas the newton (N) is related to the concept "how heavy."

These expressions, although they may seem strange at first, are already being used in quite a few large organizations which find that they cause no problems, quickly become current and eliminate continual explanation and definition. It is easy to remember that if something is expressed in kilograms, it refers to the quantity of matter, i.e. mass.

9. The Metric Technical Language

The change over to the SI presents a unique opportunity to use a correct unambiguous terminology in connection with concepts such as *mass, weight, measure, dimension*, etc. from the start.

It is recommended that the following terms be consistently used in the future. The first column gives the terms used at present. The second column gives the terms that should be used to make a clear distinction between the concepts *mass* and *weight*. Use of these terms will ensure that the units kilogram (kg) and newton (N) are applied correctly.

Existing term	New English term
A	
apothecaries' weight	apothecaries' masspiece
assize weight	assize masspiece
avoirdupois weight	avoirdupois masspiece
B	
balance weight	balance piece
basis weight	basis mass
beam scale	beam balance
	beam massmeter

box-end beam scale	box-end beam massmeter
bulk handling	bulk handling
bulk installation	bulk installation
bulk intake	bulk intake
bulk price	bulk price
bulk storage	bulk storage
bulk store	bulk store
bulk transport	bulk transport
bulk weight	bulk mass
bushel weight	mass per hectolitre

C

carat weight	carat masspiece
counterpoise	counterpoise
counter scale	counter massmeter
	counter scale
counting scale	counting scale
crane scale	crane massmeter
	crane scale
crane weigher	crane massmeter
	crane scale
cream test scale	cream test massmeter
	cream test scale
	cream test balance

D

dead-weight scale	dead-mass balance
denomination of measure	denomination of measure
dimension	dimension
drained weight	drained mass

F

fine weighing	fine mass-measuring

G

grain in bulk	grain in bulk
grain weight	grain masspiece
gross weight	gross mass

H

heavy-weight material	heavy-weight material *or* high-mass material
hopper scale	hopper massmeter hopper scale

L

lightweight material	lightweight material *or* low-mass material

M

material weight	material mass
measure	measure
measure of weight (e.g. measure of weight may be applied in selling liquids)	mass unit unit of mass (e.g. certain liquids may be sold in mass units)
measuring unit	measuring unit, unit of measurement
medium-weight material	medium-weight material *or* medium-mass material
micrometer scale	micrometer massmeter micrometer scale

N

net weight	net mass

P

person weigher	person massmeter person scale
platform scale	platform massmeter platform scale
post office beam scale	post office massmeter post office scale post office balance

S

scale	massmeter scale

short weight	short mass (expressed in kg)
sliding poise	sliding poise
spring balance 1. (calibrated in kg)	spring massmeter
	spring balance
spring balance 2. (calibrated in N)	spring weightmeter
	spring balance
standard weight 1.	standard mass
standard weight 2.	standard masspiece
steelyard	steelyard
suspended weigher	suspended massmeter
	suspended scale
swan-neck beam scale	swan-neck massmeter
	swan-neck balance
system of measures	system of measurement
system of weights and measures	system of measuring units

T

tare weight	tare
troy weight	troy masspiece

U

underweight	undermass (e.g. the bag is 10 kg undermass)
unit of measurement	unit of measurement, measuring unit
unit of weight 1.	unit of mass (the kilogram)
unit of weight 2.	unit of weight (the newton)

W

wall beam	wall massmeter
	wall scale
weigh 1. (the process of determining mass)	measure the mass
	determine the mass
weigh 2. (v.i., e.g. the object weighs 2 lb)	the mass is
	has a mass of
weighbridge	mass-measuring bridge
weighing	1. mass-measuring
	2. mass measurement

weighing balance	massmeter
	balance
weighing capacity	mass-measuring capacity
weighing instrument	massmeter
weighing process	mass-measuring process
weighing system	mass-measuring system
weigh out	measure out (e.g. measure out 5 kg)
weight 1. (quantity of matter expressed in lb, e.g. the butter has a weight of 1 lb)	mass (e.g. the butter has a mass of 500 g)
weight 2. (object of known mass used to determine the mass of other objects by using a balance, e.g. place the 1 lb weight on the scale)	masspiece (e.g. place a 500 g masspiece on the balance)
weight 3. (gravitational pull of the earth on an object expressed in newtons, e.g. the weight of a kilogram of butter is less than 2 newtons on the moon)	weight
weight 4. (object on which the weight in newtons is stamped as well as the locality where it has the weight in question, e.g. hang the 1 lb weight from the testing machine)	weightpiece (e.g. hang the 10 N weightpiece from the testing machine)
weight 5. (a general name for objects that have heaviness but are not connected with units, e.g. a paper weight, curtain weight)	weight
weight 6. (in everyday and idiomatic expressions having no connection with units, such as: pulls his weight; worth his weight in gold; an argument carries weight)	weight

weight 7. (v. to attach factors indicative of their relative frequency or importance to the various items of a frequency distribution, e.g. prices were weighted)	weight
weight change	mass change (expressed in kg)
weight class	weight class *or* mass class
weight determination 1. (in lb-mass)	mass determination (in kilograms)
weight determination 2. (in lb-force)	weight determination (in newtons)
weight lifter	weight lifter
weight limits	mass limits
weight pan	masspiece pan
weight problems	weight problems *or* mass problems
weight range	mass-measuring range
weight registering	mass-registering
weight return	mass return
weights and measures	measuring units

10. Conclusion

Because this glossary is to be used by metrication officers, translators, linguists, terminologists, educationists and other experts, it contains more terms than those normally encountered by the man in the street. This may create the impression that the whole subject is difficult. This is not the case. In everyday life it is quite sufficient to remember the following:

a) The United States is changing over to the *Metric System of Measuring Units.*
b) In this system the kilogram is the *base unit of mass.*
c) To determine mass we *measure it by means of a massmeter.*
d) The term *massmeter* includes all *instruments* capable of measuring mass, thus also *scales* and *balances,* terms that will continue to be used.
e) We place *masspieces* on a balance when measuring the mass of an object. The mass of a masspiece is constant everywhere on earth and in space and is indicated on the masspiece.

To clarify our terminology it is necessary that a specific meaning should consistently be attached to certain words and that we should not use the same word for different meanings in this context as has happened so often in the past. In this regard the following terms are of importance:

measure:
any physical object with a known denomination used to determine the value of a quantity by direct comparison. A ruler or measuring tape is a measure of length, a litre cup is a measure of volume and a masspiece is a mass measure.

measuring unit:
units such as the metre, kilogram, second, ampere. They should not be referred to as measures.

denomination of a measure:
the value or size of a measure. A masspiece of 2 kg has a denomination of 2 kg; a volume measure with a capacity of 500 m*l* has a denomination of 500 m*l*.

measurement:
the completed measuring process giving a result expressed in measuring units.

SI UNITS, MULTIPLES, SUBMULTIPLES AND PREFIXES

ampere	A	are	a
ampere metre squared	A.m²	atto-	a
ampere per metre	A/m	bar	bar
ampere per millimetre	A/mm	candela	cd
ampere per square centimetre	A/cm²	candela per square metre	cd/m²
ampere per square metre	A/m²	centi-	c
		centimetre	cm
		centipoise	cP
ampere per square millimetre	A/mm²	centistokes	cSt
		coulomb	C

coulomb metre	C.m	gram per cubic	
coulomb per cubic		centimetre	g/cm³
centimetre	C/cm³	gram per litre	g/*l*
coulomb per cubic		gram per millilitre	g/m*l*
metre	C/m³	gram per mole	g/mo*l*
coulomb per cubic		hectare	ha
millimetre	C/mm³	hecto-	h
coulomb per square		hectobar	hbar
centimetre	C/cm²	henry	H
coulomb per square		henry per metre	H/m
metre	C/m²	hertz	Hz
coulomb per square		hour	hr
millimetre	C/mm²	joule	J
cubic centimetre	cm³	joule per degree	
cubic decimetre	dm³	celsius	J/°C
cubic metre	m³	joule per kelvin	J/K
cubic millimetre	mm³	joule per kilogram	J/kg
cubic metre per		joule per kilogram	
kilomole	m³/kmol	degree celsius	J/kg.°C
cubic metre per		joule per kilogram	
mole	m³/mol	kelvin	J/kg.K
day	d	joule per kilomole	J/kmo*l*
deca-	da	joule per kilomole	
decanewton	daN	degree celsius	J/kmo*l*.°C
deci-	d	joule per kilomole	
decimetre	dm	kelvin	J/kmo*l*.K
degree (angles)	...°	joule per mole	J/mo*l*
degree celsius	°C	joule per mole	
farad	F	degree celsius	J/mo*l*.°C
farad per metre	F/m	joule per mole	
femto-	f	kelvin	J/mo*l*.K
giga-	G	joule per square	
gigahertz	GHz	centimetre	J/cm²
gigajoule	GJ	joule per square	
giganewton per		metre	J/m²
square metre	GN/m²	kelvin	K
gigaohm	GΩ	kilo-	k
gigaohm metre	GΩ.m	kiloampere	kA
gigawatt	GW	kiloampere per	
gigawatt hour	GW.h	metre	kA/m
gram	g	kiloampere per	
		square metre	kA/m²

kilocoulomb	kC
kilocoulomb per cubic metre	kC/m³
kilocoulomb per square metre	kC/m²
kilogram	kg
kilogram metre per second	kg.m/s
kilogram metre squared per second	kg.m²/s
kilogram per cubic decimetre	kg/dm³
kilogram per cubic metre	kg/m³
kilogram per litre	kg/l
kilogram per mole	kg/mol
kilohertz	kHz
kilojoule	kJ
kilojoule per degree celsius	kJ/°C
kilojoule per kelvin	kJ/K
kilojoule per kilogram	kJ/kg
kilojoule per kilogram degree celsius	kJ/kg.°C
kilojoule per kilogram kelvin	kJ/kg.K
kilojoule per square metre	kJ/m²
kilolitre	kl
kilometre	km
kilometre per hour	km/h
kilomole	kmol
kilomole per cubic metre	kmol/m³
kilomole per kilogram	kmol/kg
kilomole per litre	kmol/l
kilonewton	kN
kilonewton metre	kN.m

kilonewton per square metre	kN/m²
kilo-ohm	kΩ
kilo-ohm metre	kΩ.m
kilosecond	ks
kilosiemens	kS
kilosiemens per metre	kS/m
kilovolt	kV
kilovolt per metre	kV/m
kilowatt	kW
kilowatt hour	kW.h
kilowatt per square metre	kW/m²
kiloweber per metre	kWb/m
litre	l
lumen	lm
lumen hour	lm.h
lumen second	lm.s
lux	lx
mega-	M
mega-ampere per square metre	MA/m²
megacoulomb	MC
megacoulomb per cubic metre	MC/m³
megacoulomb per square metre	MC/m²
megagram	Mg
megagram per cubic metre	Mg/m³
megahertz	MHz
megajoule	MJ
megajoule per kilogram	MJ/kg
meganewton	MN
meganewton metre	MN.m
meganewton per square metre	MN/m²
megaohm	MΩ
megaohm metre	MΩ.m
megasiemens per metre	MS/m

megavolt	MV	milliampere	mA
megavolt per metre	MV/m	millibar	mbar
megawatt	MW	millicoulomb	mC
megawatt hour	MW.h	millifarad	mF
megawatt per square		millifarad per metre	mF/m
metre	MW/m²	milligram	mg
metre	m	millihenry	mH
metre per second	m/s	millijoule	mJ
metre per second		millilitre	m*l*
squared	m/s²	millimetre	mm
metre raised to the		millimetre squared	
fourth power	m⁴	per second	mm²/s
metre squared per		millinewton	mN
second	m²/s	millinewton per	
metric ton	t	metre	mN/m
metric ton per cubic		millinewton per	
metre	t/m³	square metre	mN/m²
micro-	μ	millinewton second	
microampere	μA	per metre	
microbar	μbar	squared	mN.s/m²
microcoulomb	μC	milliohm	mΩ
microfarad	μF	milliohm metre	mΩ.m
microfarad per metre	μF/m	milliradian	mrad
microgram	μg	millisecond	ms
microhenry	μH	millisiemens	mS
microhenry per		millitesla	mT
metre	μH/m	millivolt	mV
micrometre	μm	millivolt per metre	mV/m
micronewton	μN	milliwatt	mW
micronewton metre	μN.m	milliweber	mWb
micronewton per		minute (angles)	...′
square metre	μN/m²	minute	min
micro-ohm	μΩ	mole	mol
micro-ohm metre	μΩ.m	mole per cubic	
microradian	μrad	metre	mol/m³
microsecond	μs	mole per kilogram	mol/kg
microsiemens	μS	mole per litre	mol/*l*
microtesla	μT	nano-	n
microvolt	μV	nanoampere	nA
microvolt per metre	μV/m	nanocoulomb	nC
microwatt	μW	nanofarad	nF
milli-	m	nanofarad per metre	nF/m

nanohenry	nH	second (angles)	..."
nanohenry per metre	nH/m	siemens	S
nanometre	nm	siemens per metre	S/m
nano-ohm metre	nΩ.m	square centimetre	cm²
nanosecond	ns	square decimetre	dm²
nanotesla	nT	square kilometre	km²
nanowatt	nW	square metre	m²
newton	N	square millimetre	mm²
newton metre	N.m	steradian	sr
newton metre		tera-	T
squared per		terahertz	THz
ampere	N.m²/A	terajoule	TJ
newton per metre	N/m	terawatt	TW
newton per square		tesla	T
metre	N/m²	volt	V
newton per square		volt ampere	V.A
millimetre	N/mm²	volt per centimetre	V/cm
newton second per		volt per metre	V/m
metre squared	N.s/m²	volt per millimetre	V/mm
ohm	Ω	volt second	V.s
ohm metre	Ω.m	watt	W
ohm millimetre	Ω.mm	watt hour	W.h
pascal	Pa	watt per metre	
per degree celsius	°C⁻¹	degree celsius	W/m.°C
per henry	H⁻¹	watt per metre	
per kelvin	K⁻¹	kelvin	W/m.K
per ohm	Ω⁻¹	watt per square	
per ohm metre	Ω⁻¹.m⁻¹	metre	W/m²
per second	s⁻¹	watt per square	
pico-	p	metre degree	
picoampere	pA	celsius	W/m².°C
picocoulomb	pC	watt per square	
picofarad	pF	metre kelvin	W/m.²K
picofarad per metre	pF/m	watt per steradian	W/sr
picohenry	pH	weber	Wb
radian	rad	weber metre	Wb.m
radian per second	rad/s	weber per metre	Wb/m
radian per second		weber per milli-	
squared	rad/s²	metre	Wb/mm
second	s		

VI. TABLES OF INTERRELATION

1. UNITS OF

Units	Inches	Links	Feet	Yards	Rods
1 inch =	1	0.126 263	0.083 333 3	0.027 777 8	0.005 050 51
1 link =	7.92	1	0.66	0.22	0.04
1 foot =	12	1.515 152	1	0.333 333	0.060 606 1
1 yard =	36	4.545 45	3	1	0.181 818
1 rod =	198	25	16.5	5.5	1
1 chain =	792	100	66	22	4
1 mile =	63 360	8000	5280	1760	320
1 centimetre =	0.3937	0.049 709 60	0.032 808 33	0.010 936 111	0.001 988 384
1 metre =	39.37	4.970 960	3.280 833	1.093 611 1	0.198 838 4

2. UNITS OF

Units	Square inches	Square links	Square feet	Square yards	Square rods	Square chains
1 square inch =	1	0.015 942 3	0.006 944 44	0.000 771 605	0.000 025 507 6	0.000 001 594 23
1 square link =	62.7264	1	0.4356	0.0484	0.0016	0.0001
1 square foot =	144	2.295 684	1	0.111 111 1	0.003 673 09	0.000 229 568
1 square yard =	1296	20.6612	9	1	0.033 057 85	0.002 066 12
1 square rod =	39 204	625	272.25	30.25	1	0.0625
1 square chain =	627 264	10 000	4356	484	16	1
1 acre =	6 272 640	100 000	43 560	4840	160	10
1 square mile =	4 014 489 600	64 000 000	27 878 400	3 097 600	102 400	6400
1 square centimetre =	0.154 999 69	0.002 471 04	0.001 076 387	0.000 119 598 5	0.000 003 953 67	0.000 000 247 104
1 square metre =	1549.9969	24.7104	10.763 87	1.195 985	0.039 536 7	0.002 471 04
1 hectare =	15 499 969	247 104	107 638.7	11 959.85	395.367	24.7104

3. UNITS OF

Units	Cubic inches	Cubic feet	Cubic yards
1 cubic inch =	1	0.000 578 704	0.000 021 433 47
1 cubic foot =	1728	1	0.037 037 0
1 cubic yard =	46 656	27	1
1 cubic centimetre =	0.061 023 38	0.000 035 314 45	0.000 001 307 94
1 cubic decimetre =	61.023 38	0.035 314 45	0.001 307 943
1 cubic metre =	61 023.38	35.314 45	1.307 942 8

4. UNITS OF CAPACITY

Units	Minims	Fluid drams	Fluid ounces	Gills	Liquid pints
1 minim =	1	0.016 666 7	0.002 083 33	0.000 520 833	0.000 130 208
1 fluid dram =	60	1	0.125	0.031 25	0.007 812 5
1 fluid ounce =	480	8	1	0.25	0.0625
1 gill =	1920	32	4	1	0.25
1 liquid pint =	7680	128	16	4	1
1 liquid quart =	15 360	256	32	8	2
1 gallon =	61 440	1024	128	32	8
1 millilitres =	16.2311	0.270 518	0.033 814 7	0.008 453 68	0.002 113 42
1 liter =	16 231.1	270.518	33.8147	8.453 68	2.113 42
1 cubic inch =	265.974	4.432 90	0.554 113	0.138 528	0.034 632 0

OF UNITS OF MEASUREMENT

LENGTH

Chains	Miles	Centimetres	Metres		Units
0.001 262 63	0.000 015 782 8	2.540 005	0.025 400 05		=1 inch
0.01	0.000 125	20.116 84	0.201 168 4		=1 link
0.015 151 5	0.000 189 393 9	30.480 06	0.304 800 6		=1 foot
0.045 454 5	0.000 568 182	91.440 18	0.914 401 8		=1 yard
0.25	0.003 125	502.9210	5.029 210		=1 rod
1	0.0125	2011.684	20.116 84		=1 chain
80	1	160 934.72	1609.3472		=1 mile
0.000 497 096 0	0.000 006 213 699	1	0.01		=1 centimetre
0.049 709 60	0.000 621 369 9	100	1		=1 metre

AREA

Acres	Square miles	Square centimetres	Square metres	Hectares	Units
0.000 000 159 423	0.000 000 000 249 1	6.451 626	0.000 645 162 6	0.000 000 064 516	=1 square inch
0.000 01	0.000 000 015 625	404.6873	0.040 468 73	0.000 004 046 87	=1 square link
0.000 022 956 8	0.000 000 035 870 1	929.0341	0.092 903 41	0.000 000 290 34	=1 square foot
0.000 206 612	0.000 000 322 831	8361.307	0.836 130 7	0.000 083 613 1	=1 square yard
0.006 25	0.000 009 765 625	252 929.5	25.292 95	0.002 529 295	=1 square rod
0.1	0.000 156 25	4 046 873	404.6873	0.040 468 7	=1 square chain
1	0.001 562 5	40 468 726	4046.873	0.404 687	=1 acre
640	1	25 899 984 703	2 589 998	258.9998	=1 square mile
0.000 000 024 710 4	0.000 000 000 038 610 06	1	0.0001	0.000 000 01	=1 square centimetre
0.000 247 104	0.000 000 386 100 6	10 000	1	0.0001	=1 square metre
2.471 04	0.003 861 006	100 000 000	10 000	1	=1 hectare

VOLUME

Cubic centimetres	Cubic decimetres	Cubic metres	Units
16.387 162	0.016 387 16	0.000 016 387 16	=1 cubic inch
28 317.016	28.317 016	0.028 317 016	=1 cubic foot
764 559.4	764.5594	0.764 559 4	=1 cubic yard
1	0.001	0.000 001	=1 cubic centimetre
1 000	1	0.001	=1 cubic decimetre
1 000 000	1000	1	=1 cubic metre

LIQUID MEASURE

Liquid quarts	Gallons	Millilitres	Liters	Cubic inches	Units
0.000 065 104	0.000 016 276	0.061 610 2	0.000 061 610 2	0.003 759 77	=1 minim
0.003 906 25	0.000 976 562	3.696 61	0.003 696 61	0.225 586	=1 fluid dram
0.031 25	0.007 812 5	29.5729	0.029 572 9	1.804 69	=1 fluid ounce
0.125	0.031 25	118.292	0.118 292	7.218 75	=1 gill
0.5	0.125	473.167	0.473 167	28.875	=1 liquid pint
1	0.25	946.333	0.946 333	57.75	=1 liquid quart
4	1	3785.332	3.785 332	231	=1 gallon
0.001 056 71	0.000 264 178	1	0.001	0.061 025 0	=1 milliliter
1.056 71	0.264 178	1000	1	61.0250	=1 liter
0.017 316 0	0.004 329 00	16.3867	0.016 386 7	1	=1 cubic inch

VI. TABLES OF INTERRELATION
5. UNITS OF CAPACITY

Units	Dry pints	Dry quarts	Pecks	Bushels
1 dry pint =	1	0.5	0.0625	0.015 625
1 dry quart =	2	1	0.125	0.031 25
1 peck =	16	8	1	0.25
1 bushel =	64	32	4	1
1 liter =	1.816 20	0.908 102	0.113 513	0.028 378
1 dekalitre =	18.1620	9.081 02	1.135 13	0.283 78
1 cubic inch=	0.029 761 6	0.014 880 8	0.001 860 10	0.000 465 025

6. UNITS OF MASS LESS

Units *	Grains	Apothecaries' scruples	Pennyweights	Avoirdupois drams	Apothecaries' drams	Avoirdupois ounces
1 grain =	1	0.05	0.041 666 67	0.036 571 43	0.016 666 7	0.002 285 71
1 apoth. scruple =	20	1	0.833 333 3	0.731 428 6	0.333 333	0.045 714 3
1 pennyweight =	24	1.2	1	0.877 714 3	0.4	0.054 857 1
1 avoir. dram =	27.343 75	1.367 187 5	1.139 323	1	0.455 729 2	0.0625
1 apoth. dram =	60	3	2.5	2.194 286	1	0.137 142 9
1 avoir. ounce =	437.5	21.875	18.229 17	16	7.291 67	1
1 apoth. or troy ounce=	480	24	20	17.554 28	8	1.097 142 9
1 apoth. or troy pound=	5760	288	240	210.6514	96	13.165 714
1 avoir. pound =	7000	350	291.6667	256	116.6667	16
1 milligram =	0.015 432 356	0.000 771 618	0.000 643 014 8	0.000 564 383 3	0.000 257 205 9	0.000 035 273 96
1 gram =	15.432 356	0.771 618	0.643 014 85	0.564 383 3	0.257 205 9	0.035 273 96
1 kilogram =	15 432.356	771.6178	643.014 85	564.383 32	257.205 94	35.273 96

7. UNITS OF MASS

Units	Avoirdupois ounces	Avoirdupois pounds	Short hundred-weights	Short tons
1 avoirdupois ounce =	1	0.0625	0.000 625	0.000 031 25
1 avoirdupois pound =	16	1	0.01	0.0005
1 short hundredweight=	1600	100	1	0.05
1 short ton =	32 000	2000	20	1
1 long ton =	35 840	2240	22.4	1.12
1 kilogram =	35.273 957	2.204 622 34	0.022 046 223	0.001 102 311.2
1 metric ton =	35 273.957	2204.622 34	22.046 223	1.102 311 2

* "Avoir," is now abbreviated "avdp".

OF UNITS OF MEASUREMENT
DRY MEASURE

Litres	Dekalitres	Cubic inches	Units
0.550 599	0.055 060	33.600 312 5	=1 dry pint
1.101 198	0.110 120	67.200 625	=1 dry quart
8.809 58	0.880 958	537.605	=1 peck
35.2383	3.523 83	2150.42	=1 bushel
1	0.1	61.0250	=1 liter
10	1	610.250	=1 dekalitre
0.016 386 7	0.001 638 67	1	=1 cubic inch

THAN POUNDS AND KILOGRAMS

Apothecaries' or troy ounces	Apothecaries' or troy pounds	Avoirdupois pounds	Milligrams	Grams	Kilograms	Units
0.002 083 33	0.000 173 611 1	0.000 142 857 1	64.798 918	0.064 798 918	0.000 064 798 9	=1 grain
0.041 666 7	0.003 472 222	0.002 857 143	1295.9784	1.295 978 4	0.001 295 978	=1 apoth. scruple
0.05	0.004 166 667	0.003 428 571	1555.1740	1.555 174 0	0.001 555 174	=1 pennyweight
0.056 966 146	0.004 747 178 8	0.003 906 25	1771.8454	1.771 845 4	0.001 771 845	=1 avoir. dram
0.125	0.010 416 667	0.008 571 429	3887.9351	3.887 935 1	0.003 887 935	=1 apoth. dram
0.911 458 3	0.075 954 861	0.0625	28 349.527	28.349 527	0.028 349 53	=1 avoir. ounce
1	0.083 333 33	0.068 571 43	31 103.481	31.103 481	0.031 103 48	=1 apoth. or troy ounce
12	1	0.822 857 1	373 241.77	373.241 77	0.373 241 77	=1 apoth. or troy pound
14.583 333	1.215 277 8	1	453 592.4277	453.592 4277	0.453 592 427 7	=1 avoir. pound
0.000 032 150 74	0.000 002 679 23	0.000 002 204 62	1	0.001	0.000 001	=1 milligram
0.032 150 74	0.002 679 23	0.002 204 62	1000	1	0.001	=1 gram
32.150 742	2.679 228 5	2.204 622 341	1 000 000	1000	1	=1 kilogram

GREATER THAN AVOIRDUPOIS OUNCES

Long tons	Kilograms	Metric tons	Units
0.000 027 901 79	0.028 349 53	0.000 028 349 53	=1 avoirdupois ounce
0.000 446 428 6	0.453 592 427 7	0.000 453 592 43	=1 avoirdupois pound
0.041 642 86	45.359 243	0.045 359 243	=1 short hundredweight
0.892 857 1	907.184 86	0.907 184 86	=1 short ton
1	1016.047 04	1.016 047 04	=1 long ton
0.000 984 206 4	1	0.001	=1 kilogram
0.984 206 40	1000	1	=1 metric ton

1. LENGTH

Inches (in.)	Millimetres (mm)	Feet (ft)	Metres (m)	Yards (yd)	Metres (m)	Rods (rd)	Metres (m)	U.S. miles (mi)	Kilometres (km)
1=	25.4001	1=	0.304 801	1=	0.914 402	1=	5.029 21	1=	1.609 347
2=	50.8001	2=	0.609 601	2=	1.828 804	2=	10.058 42	2=	3.218 694
3=	76.2002	3=	0.914 402	3=	2.743 205	3=	15.087 63	3=	4.828 042
4=	101.6002	4=	1.219 202	4=	3.657 607	4=	20.116 84	4=	6.437 389
5=	127.0003	5=	1.524 003	5=	4.572 009	5=	25.146 05	5=	8.046 736
6=	152.4003	6=	1.828 804	6=	5.486 411	6=	30.175 26	6=	9.656 083
7=	177.8004	7=	2.133 604	7=	6.400 813	7=	35.204 47	7=	11.265 431
8=	203.2004	8=	2.438 405	8=	7.315 215	8=	40.233 68	8=	12.874 778
9=	228.6005	9=	2.743 205	9=	8.229 616	9=	45.262 89	9=	14.484 125
0.039 37=1		3.280 83=1		1.093 611=1		0.198 838=1		0.621 370=1	
0.078 74=2		6.561 67=2		2.187 222=2		0.397 677=2		1.242 740=2	
0.118 11=3		9.842 50=3		3.280 833=3		0.596 515=3		1.864 110=3	
0.157 48=4		13.123 33=4		4.374 444=4		0.795 354=4		2.485 480=4	
0.196 85=5		16.404 17=5		5.468 056=5		0.994 192=5		3.106 850=5	
0.236 22=6		19.685 00=6		6.561 667=6		1.193 030=6		3.728 220=6	
0.275 59=7		22.965 83=7		7.655 278=7		1.391 869=7		4.349 590=7	
0.314 96=8		26.246 67=8		8.748 889=8		1.590 707=8		4.970 960=8	
0.354 33=9		29.527 50=9		9.842 500=9		1.789 545=9		5.592 330=9	

2. AREA

Square inches (sq in.)	Square centimetres (cm²)	Square feet (sq ft)	Square metres (m²)	Square yards (sq yd)	Square metres (m²)	Acres (acre)	Hectares (ha)	Square miles (sq mi)	Square Kilometres (km²)
1=	6.452	1=	0.092 90	1=	0.8361	1=	0.4047	1=	2.5900
2=	12.903	2=	0.185 81	2=	1.6723	2=	0.8094	2=	5.1800
3=	19.355	3=	0.278 71	3=	2.5084	3=	1.2141	3=	7.7700
4=	25.807	4=	0.371 61	4=	3.3445	4=	1.6187	4=	10.3600
5=	32.258	5=	0.464 52	5=	4.1807	5=	2.0234	5=	12.9500
6=	38.710	6=	0.557 42	6=	5.0168	6=	2.4281	6=	15.5400
7=	45.161	7=	0.650 32	7=	5.8529	7=	2.8328	7=	18.1300
8=	51.613	8=	0.743 23	8=	6.6890	8=	3.2375	8=	20.7200
9=	58.065	9=	0.836 13	9=	7.5252	9=	3.6422	9=	23.3100
0.155 00=1		10.764=1		1.1960=1		2.471=1		0.3861=1	
0.310 00=2		21.528=2		2.3920=2		4.942=2		0.7722=2	
0.465 00=3		32.292=3		3.5880=3		7.413=3		1.1583=3	
0.620 00=4		43.055=4		4.7839=4		9.884=4		1.5444=4	
0.775 00=5		53.819=5		5.9799=5		12.355=5		1.9305=5	
0.930 00=6		64.583=6		7.1759=6		14.826=6		2.3166=6	
1.085 00=7		75.347=7		8.3719=7		17.297=7		2.7027=7	
1.240 00=8		86.111=8		9.5679=8		19.768=8		3.0888=8	
1.395 00=9		96.875=9		10.7639=9		22.239=9		3.4749=9	

3. VOLUME

Cubic inches (cu in.)	Cubic centimetres (cm³)	Cubic feet (cu ft)	Cubic metres (m³)	Cubic yards (cu yd)	Cubic metres (m³)	Cubic inches (cu in.)	Litres (litre)	Cubic feet (cu ft)	Litres (litre)
1=	16.3872	1=	0.028 317	1=	0.7646	1=	0.016 386 7	1=	28.316
2=	32.7743	2=	0.056 634	2=	1.5291	2=	0.032 773 4	2=	56.633
3=	49.1615	3=	0.084 951	3=	2.2937	3=	0.049 160 2	3=	84.949
4=	65.5486	4=	0.113 268	4=	3.0582	4=	0.065 546 9	4=	113.265
5=	81.9358	5=	0.141 585	5=	3.8228	5=	0.081 933 6	5=	141.581
6=	98.3230	6=	0.169 902	6=	4.5874	6=	0.098 320 3	6=	169.898
7=	114.7101	7=	0.198 219	7=	5.3519	7=	0.114 707 0	7=	198.214
8=	131.0973	8=	0.226 536	8=	6.1165	8=	0.131 093 8	8=	226.530
9=	147.4845	9=	0.254 853	9=	6.8810	9=	0.147 480 5	9=	254.846
0.061 02=1		35.314=1		1.3079=1		61.025=1		0.035 315=1	
0.122 05=2		70.629=2		2.6159=2		122.050=2		0.070 631=2	
0.183 07=3		105.943=3		3.9238=3		183.075=3		0.105 946=3	
0.244 09=4		141.258=4		5.2318=4		244.100=4		0.141 262=4	
0.305 12=5		176.572=5		6.5397=5		305.125=5		0.176 577=5	
0.366 14=6		211.887=6		7.8477=6		366.150=6		0.211 892=6	
0.427 16=7		247.201=7		9.1556=7		427.175=7		0.247 208=7	
0.488 19=8		282.516=8		10.4635=8		488.200=8		0.282 523=8	
0.549 21=9		317.830=9		11.7715=9		549.225=9		0.317 839=9	

4. CAPACITY—LIQUID MEASURE

U. S. fluid drams (fl dr) = Millilitres (ml)	U. S. fluid ounces (fl oz) = Millilitres (ml)	U. S. liquid pints (pt) = Litres (litre)	U. S. liquid quarts (qt) = Litres (litre)	U. S. gallons (gal) = Litres (litre)
1= 3.6966	1= 29.573	1=0.473 17	1=0.946 33	1= 3.785 33
2= 7.3932	2= 59.146	2=0.946 33	2=1.892 67	2= 7.570 66
3=11.0898	3= 88.719	3=1.419 50	3=2.839 00	3=11.356 00
4=14.7865	4=118.292	4=1.892 67	4=3.785 33	4=15.141 33
5=18.4831	5=147.865	5=2.365 83	5=4.731 67	5=18.926 66
6=22.1797	6=177.437	6=2.839 00	6=5.678 00	6=22.711 99
7=25.8763	7=207.010	7=3.312 17	7=6.624 33	7=26.497 33
8=29.5729	8=236.583	8=3.785 33	8=7.570 66	8=30.282 66
9=33.2695	9=266.156	9=4.258 50	9=8.517 00	9=34.067 99
0.270 52=1	0.033 815=1	2.1134=1	1.056 71=1	0.264 18=1
0.541 04=2	0.067 629=2	4.2268=2	2.113 42=2	0.528 36=2
0.811 55=3	0.101 444=3	6.3403=3	3.170 13=3	0.792 53=3
1.082 07=4	0.135 259=4	8.4537=4	4.226 84=4	1.056 71=4
1.352 59=5	0.169 074=5	10.5671=5	5.283 55=5	1.320 89=5
1.623 11=6	0.202 888=6	12.6805=6	6.340 26=6	1.585 07=6
1.893 63=7	0.236 703=7	14.7939=7	7.396 97=7	1.849 24=7
2.164 14=8	0.270 518=8	16.9074=8	8.453 68=8	2.113 42=8
2.434 66=9	0.304 333=9	19.0208=9	9.510 39=9	2.377 60=9

5. CAPACITY—DRY MEASURE

U. S. dry quarts (qt) = Litres (litre)	U. S. pecks (pk) = Litres (litre)	U. S. pecks (pk) = Dekalitres (dkl)	U. S. bushels (bu) = Hectolitres (hl)	U. S. bushels per acre = Hectolitres per hectare
1=1.1012	1= 8.810	1=0.8810	1=0.352 38	1=0.8708
2=2.2024	2=17.619	2=1.7619	2=0.704 77	2=1.7415
3=3.3036	3=26.429	3=2.6429	3=1.057 15	3=2.6123
4=4.4048	4=35.238	4=3.5238	4=1.409 53	4=3.4830
5=5.5060	5=44.048	5=4.4048	5=1.761 92	5=4.3538
6=6.6072	6=52.857	6=5.2857	6=2.114 30	6=5.2245
7=7.7084	7=61.667	7=6.1667	7=2.466 68	7=6.0953
8=8.8096	8=70.477	8=7.0477	8=2.819 07	8=6.9660
9=9.9108	9=79.286	9=7.9286	9=3.171 45	9=7.8368
0.9081=1	0.113 51=1	1.1351=1	2.8378=1	1.1484=1
1.8162=2	0.227 03=2	2.2703=2	5.6756=2	2.2969=2
2.7243=3	0.340 54=3	3.4054=3	8.5135=3	3.4453=3
3.6324=4	0.454 05=4	4.5405=4	11.3513=4	4.5937=4
4.5405=5	0.567 56=5	5.6756=5	14.1891=5	5.7421=5
5.4486=6	0.681 08=6	6.8108=6	17.0269=6	6.8906=6
6.3567=7	0.794 59=7	7.9459=7	19.8647=7	8.0390=7
7.2648=8	0.908 10=8	9.0810=8	22.7026=8	9.1874=8
8.1729=9	1.021 61=9	10.2161=9	25.5404=9	10.3359=9

6. MASS

Grains (grain) = Grams (g)	Apothecaries' drams (dr ap or ℨ) = Grams (g)	Troy ounces (oz t) = Grams (g)	Avoirdupois ounces (oz avdp) = Grams (g)	Avoirdupois pounds (lb avdp) = Kilograms (kg)
1=0.064 799	1= 3.8879	1= 31.103	1= 28.350	1=0.453 59
2=0.129 598	2= 7.7759	2= 62.207	2= 56.699	2=0.907 18
3=0.194 397	3=11.6638	3= 93.310	3= 85.049	3=1.360 78
4=0.259 196	4=15.5517	4=124.414	4=113.398	4=1.814 37
5=0.323 995	5=19.4397	5=155.517	5=141.748	5=2.267 96
6=0.388 794	6=23.3276	6=186.621	6=170.097	6=2.721 55
7=0.453 592	7=27.2155	7=217.724	7=198.447	7=3.175 15
8=0.518 391	8=31.1035	8=248.828	8=226.796	8=3.628 74
9=0.583 190	9=34.9914	9=279.931	9=255.146	9=4.082 33
15.4324=1	0.257 21=1	0.032 151=1	0.035 274=1	2.204 62=1
30.8647=2	0.514 41=2	0.064 301=2	0.070 548=2	4.409 24=2
46.2971=3	0.771 62=3	0.096 452=3	0.105 822=3	6.613 87=3
61.7294=4	1.028 82=4	0.128 603=4	0.141 096=4	8.818 49=4
77.1618=5	1.286 03=5	0.160 754=5	0.176 370=5	11.023 11=5
92.5941=6	1.543 24=6	0.192 904=6	0.211 644=6	13.227 73=6
108.0265=7	1.800 44=7	0.225 055=7	0.246 918=7	15.432 36=7
123.4589=8	2.057 65=8	0.257 206=8	0.282 192=8	17.636 98=8
138.8912=9	2.314 85=9	0.289 357=9	0.317 466=9	19.841 60=9

COMPARISON OF THE VARIOUS TONS AND POUNDS IN USE IN THE UNITED STATES (FROM 1 TO 9 UNITS)

Troy pounds	Avoirdupois pounds	Kilograms	Short tons	Long tons	Metric tons
1	0.822 857	0.373 24	0.000 411 43	0.000 367 35	0.000 373 24
2	1.645 71	0.746 48	0.000 822 86	0.000 734 69	0.000 746 48
3	2.468 57	1.119 73	0.001 234 29	0.001 102 04	0.001 119 73
4	3.291 43	1.492 97	0.001 645 71	0.001 469 39	0.001 492 97
5	4.114 29	1.866 21	0.002 057 14	0.001 836 73	0.001 866 21
6	4.937 14	2.239 45	0.002 468 57	0.002 204 08	0.002 239 45
7	5.760 00	2.612 69	0.002 880 00	0.002 571 43	0.002 612 69
8	6.582 86	2.985 93	0.003 291 43	0.002 938 78	0.002 985 93
9	7.405 71	3.359 18	0.003 702 86	0.003 306 12	0.003 359 18
1.215 28	1	0.453 59	0.0005	0.000 446 43	0.000 453 59
2.430 56	2	0.907 18	0.0010	0.000 892 86	0.000 907 18
3.645 83	3	1.360 78	0.0015	0.001 339 29	0.001 360 78
4.861 11	4	1.814 37	0.0020	0.001 785 71	0.001 814 37
6.076 39	5	2.267 96	0.0025	0.002 232 14	0.002 267 96
7.291 67	6	2.721 55	0.0030	0.002 678 57	0.002 721 55
8.506 94	7	3.175 15	0.0035	0.003 125 00	0.003 175 15
9.722 22	8	3.628 74	0.0040	0.003 571 43	0.003 628 74
10.937 50	9	4.082 33	0.0045	0.004 017 86	0.004 082 33
2.679 23	2.204 62	1	0.001 102 31	0.000 984 21	0.001
5.358 46	4.409 24	2	0.002 204 62	0.001 968 41	0.002
8.037 69	6.613 87	3	0.003 306 93	0.002 952 62	0.003
10.716 91	8.818 49	4	0.004 409 24	0.003 936 83	0.004
13.396 14	11.023 11	5	0.005 511 56	0.004 921 03	0.005
16.075 37	13.227 73	6	0.006 613 87	0.005 905 24	0.006
18.754 60	15.432 36	7	0.007 716 18	0.006 889 44	0.007
21.433 83	17.636 98	8	0.008 818 49	0.007 873 65	0.008
24.113 06	19.841 60	9	0.009 920 80	0.008 857 86	0.009

Troy pounds	Avoirdupois pounds	Kilograms	Short tons	Long tons	Metric tons
2430.56	2000	907.18	1	0.892 86	0.907 18
4861.11	4000	1814.37	2	1.785 71	1.814 37
7291.67	6000	2721.55	3	2.678 57	2.721 55
9722.22	8000	3628.74	4	3.571 43	3.628 74
12 152.78	10 000	4535.92	5	4.464 29	4.535 92
14 583.33	12 000	5443.11	6	5.357 14	5.443 11
17 013.89	14 000	6350.29	7	6.250 00	6.350 29
19 444.44	16 000	7257.48	8	7.142 86	7.257 48
21 875.00	18 000	8164.66	9	8.035 71	8.164 66
2722.22	2240	1016.05	1.12	1	1.016 05
5444.44	4480	2032.09	2.24	2	2.032 09
8166.67	6720	3048.14	3.36	3	3.048 14
10 888.89	8960	4064.19	4.48	4	4.064 19
13 611.11	11 200	5080.24	5.60	5	5.080 24
16 333.33	13 440	6096.28	6.72	6	6.096 28
19 055.56	15 680	7112.32	7.84	7	7.112 32
21 777.78	17 920	8128.38	8.96	8	8.128 38
24 500.00	20 160	9144.42	10.08	9	9.144 42
2679.23	2204.62	1000	1.102 31	0.984 21	1
5358.46	4409.24	2000	2.204 62	1.968 41	2
8037.69	6613.87	3000	3.306 93	2.952 62	3
10 716.91	8818.49	4000	4.409 24	3.936 83	4
13 396.14	11 023.11	5000	5.511 56	4.921 03	5
16 075.37	13 227.73	6000	6.613 87	5.905 24	6
18 754.60	15 432.36	7000	7.716 18	6.889 44	7
21 433.83	17 636.98	8000	8.818 49	7.873 65	8
24 113.06	19 841.60	9000	9.920 80	8.857 86	9

VIII. SPECIAL TABLES

LENGTH—INCHES AND MILLIMETRES—EQUIVALENTS OF DECIMAL AND BINARY FRACTIONS OF AN INCH IN MILLIMETRES

From 1/64 to 1 Inch

½'s	¼'s	8ths	16ths	32ds	64ths	Milli-metres	Decimals of an inch
					1	= 0.397	0.015625
				1	2	= .794	.03125
					3	= 1.191	.046875
			1	2	4	= 1.588	.0625
					5	= 1.984	.078125
				3	6	= 2.381	.09375
					7	= 2.778	.109375
		1	2	4	8	= 3.175	.1250
					9	= 3.572	.140625
				5	10	= 3.969	.15625
					11	= 4.366	.171875
			3	6	12	= 4.763	.1875
					13	= 5.159	.203125
				7	14	= 5.556	.21875
					15	= 5.953	.234375
	1	2	4	8	16	= 6.350	.2500
					17	= 6.747	.265625
				9	18	= 7.144	.28125
					19	= 7.541	.296875
			5	10	20	= 7.938	.3125
					21	= 8.334	.328125
				11	22	= 8.731	.34375
					23	= 9.128	.359375
		3	6	12	24	= 9.525	.3750
					25	= 9.922	.390625
				13	26	= 10.319	.40625
					27	= 10.716	.421875
			7	14	28	= 11.113	.4375
					29	= 11.509	.453125
				15	30	= 11.906	.46875
					31	= 12.303	.484375
1	2	4	8	16	32	= 12.700	.5

Inch	½'s	¼'s	8ths	16ths	32ds	64ths	Milli-metres	Decimals of an inch
						33	= 13.097	0.515625
					17	34	= 13.494	.53125
						35	= 13.891	.546875
				9	18	36	= 14.288	.5625
						37	= 14.684	.578125
					19	38	= 15.081	.59375
						39	= 15.478	.609375
			5	10	20	40	= 15.875	.625
						41	= 16.272	.640625
					21	42	= 16.669	.65625
						43	= 17.066	.671875
				11	22	44	= 17.463	.6875
						45	= 17.859	.703125
		3			23	46	= 18.256	.71875
						47	= 18.653	.734375
			6	12	24	48	= 19.050	.75
						49	= 19.447	.765625
					25	50	= 19.844	.78125
						51	= 20.241	.796875
				13	26	52	= 20.638	.8125
						53	= 21.034	.828125
					27	54	= 21.431	.84375
						55	= 21.828	.859375
			7	14	28	56	= 22.225	.875
						57	= 22.622	.890625
					29	58	= 23.019	.90625
						59	= 23.416	.921875
				15	30	60	= 23.813	.9375
						61	= 24.209	.953125
					31	62	= 24.606	.96875
						63	= 25.003	.984375
1	2	4	8	16	32	64	= 25.400	1.000

VIII. SPECIAL TABLES

LENGTH—HUNDREDTHS OF AN INCH TO MILLIMETRES

From 1 to 99 Hundredths

Hundredths of an inch	0	1	2	3	4	5	6	7	8	9
	0	0.254	0.508	0.762	1.016	1.270	1.524	1.778	2.032	2.286
10	2.540	2.794	3.048	3.302	3.556	3.810	4.064	4.318	4.572	4.826
20	5.080	5.334	5.588	5.842	6.096	6.350	6.604	6.858	7.112	7.366
30	7.620	7.874	8.128	8.382	8.636	8.890	9.144	9.398	9.652	9.906
40	10.160	10.414	10.668	10.922	11.176	11.430	11.684	11.938	12.192	12.446
50	12.700	12.954	13.208	13.462	13.716	13.970	14.224	14.478	14.732	14.986
60	15.240	15.494	15.748	16.002	16.256	16.510	16.764	17.018	17.272	17.526
70	17.780	18.034	18.288	18.542	18.796	19.050	19.304	19.558	19.812	20.066
80	20.320	20.574	20.828	21.082	21.336	21.590	21.844	22.098	22.352	22.606
90	22.860	23.114	23.368	23.622	23.876	24.130	24.384	24.638	24.892	25.146

LENGTH—MILLIMETRES TO DECIMALS OF AN INCH

From 1 to 99 Units

Milli-metres	0	1	2	3	4	5	6	7	8	9
	0	0.03937	0.07874	0.11811	0.15748	0.19685	0.23622	0.27559	0.31496	0.35433
10	0.39370	.43307	.47244	.51181	.55118	.59055	.62992	.66929	.70866	.74803
20	.78740	.82677	.86614	.90551	.94488	.98425	1.02362	1.06299	1.10236	1.14173
30	1.18110	1.22047	1.25984	1.29921	1.33858	1.37795	1.41732	1.45669	1.49606	1.53543
40	1.57480	1.61417	1.65354	1.69291	1.73228	1.77165	1.81102	1.85039	1.88976	1.92913
50	1.96850	2.00787	2.04724	2.08661	2.12598	2.16535	2.20472	2.24409	2.28346	2.32283
60	2.36220	2.40157	2.44094	2.48031	2.51968	2.55905	2.59842	2.63779	2.67716	2.71653
70	2.75590	2.79527	2.83464	2.87401	2.91338	2.95275	2.99212	3.03149	3.07086	3.11023
80	3.14960	3.18897	3.22834	3.26771	3.30708	3.34645	3.38582	3.42519	3.45456	3.50393
90	3.54330	3.58267	3.62204	3.66141	3.70078	3.74015	3.77952	3.81889	3.85826	3.89763

71

LENGTH—UNITED STATES NAUTICAL MILES, INTERNATIONAL NAUTICAL MILES, AND KILOMETRES

Basic relations 1 U.S. nautical mile = 1.853 248 kilometres. 1 U.S. nautical mile = 1.000 673 9 int. nautical miles. 1 International nautical mile = 1.852 kilometres.

U.S. nautical miles	Int. nautical miles	Kilometres
0		
1	1.0007	1.8532
2	2.0013	3.7065
3	3.0020	5.5597
4	4.0027	7.4130
5	5.0034	9.2662
6	6.0040	11.1195
7	7.0047	12.9727
8	8.0054	14.8260
9	9.0061	16.6792
10	10.0067	18.5325
11	11.0074	20.3857
12	12.0081	22.2390
13	13.0088	24.0922
14	14.0094	25.9455
15	15.0101	27.7987
16	16.0108	29.6520
17	17.0115	31.5052
18	18.0121	33.3585
19	19.0128	35.2117
20	20.0135	37.0650
21	21.0142	38.9182
22	22.0148	40.7715
23	23.0155	42.6247
24	24.0162	44.4780
25	25.0168	46.3312
26	26.0175	48.1844
27	27.0182	50.0377
28	28.0189	51.8909
29	29.0195	53.7442
30	30.0202	55.5974
31	31.0209	57.4507
32	32.0216	59.3039
33	33.0222	61.1572
34	34.0229	63.0104
35	35.0236	64.8637
36	36.0243	66.7169
37	37.0249	68.5702
38	38.0256	70.4234
39	39.0263	72.2767
40	40.0270	74.1299
41	41.0276	75.9832
42	42.0283	77.8364
43	43.0290	79.6897
44	44.0297	81.5429
45	45.0303	83.3962
46	46.0310	85.2494
47	47.0317	87.1027
48	48.0323	88.9559
49	49.0330	90.8092

Int. nautical miles	U.S. nautical miles	Kilometres
0		
1	0.9993	1.8520
2	1.9987	3.7040
3	2.9980	5.5560
4	3.9973	7.4080
5	4.9966	9.2600
6	5.9960	11.1120
7	6.9953	12.9640
8	7.9946	14.8160
9	8.9939	16.6680
10	9.9933	18.5200
11	10.9926	20.3720
12	11.9919	22.2240
13	12.9912	24.0760
14	13.9906	25.9280
15	14.9899	27.7800
16	15.9892	29.6320
17	16.9886	31.4840
18	17.9879	33.3360
19	18.9872	35.1880
20	19.9865	37.0400
21	20.9859	38.8920
22	21.9852	40.7440
23	22.9845	42.5960
24	23.9838	44.4480
25	24.9832	46.3000
26	25.9825	48.1520
27	26.9818	50.0040
28	27.9811	51.8560
29	28.9805	53.7080
30	29.9798	55.5600
31	30.9791	57.4120
32	31.9785	59.2640
33	32.9778	61.1160
34	33.9771	62.9680
35	34.9764	64.8200
36	35.9758	66.6720
37	36.9751	68.5240
38	37.9744	70.3760
39	38.9737	72.2280
40	39.9731	74.0800
41	40.9724	75.9320
42	41.9717	77.7840
43	42.9710	79.6360
44	43.9704	81.4880
45	44.9697	83.3400
46	45.9690	85.1920
47	46.9683	87.0440
48	47.9677	88.8960
49	48.9670	90.7480

Kilometres	U.S. nautical	Int. nautical
0		
1	0.5396	0.5400
2	1.0792	1.0799
3	1.6188	1.6199
4	2.1584	2.1598
5	2.6980	2.6998
6	3.2376	3.2397
7	3.7772	3.7797
8	4.3167	4.3197
9	4.8563	4.8596
10	5.3959	5.3996
11	5.9355	5.9395
12	6.4751	6.4795
13	7.0147	7.0194
14	7.5543	7.5594
15	8.0939	8.0994
16	8.6335	8.6393
17	9.1731	9.1793
18	9.7127	9.7192
19	10.2523	10.2592
20	10.7919	10.7991
21	11.3315	11.3391
22	11.8711	11.8790
23	12.4106	12.4190
24	12.9502	12.9590
25	13.4898	13.4989
26	14.0294	14.0389
27	14.5690	14.5788
28	15.1086	15.1188
29	15.6482	15.6587
30	16.1878	16.1987
31	16.7274	16.7387
32	17.2670	17.2786
33	17.8066	17.8186
34	18.3462	18.3585
35	18.8858	18.8985
36	19.4254	19.4384
37	19.9649	19.9784
38	20.5045	20.5184
39	21.0441	21.0583
40	21.5837	21.5983
41	22.1233	22.1382
42	22.6629	22.6782
43	23.2025	23.2181
44	23.7421	23.7581
45	24.2817	24.2981
46	24.8213	24.8380
47	25.3609	25.3780
48	25.9005	25.9179
49	26.4401	26.4579

	A	B	C	
50	26.9978	26.9797	92.6000	50
1	27.5378	27.5193	94.4520	1
2	28.0778	28.0588	96.3040	2
3	28.6177	28.5984	98.1560	3
4	29.1577	29.1380	100.0080	4
5	29.6976	29.6776	101.8600	5
6	30.2376	30.2172	103.7120	6
7	30.7775	30.7568	105.5640	7
8	31.3175	31.2964	107.4160	8
9	31.8575	31.8360	109.2680	9
60	32.3974	32.3756	111.1200	60
1	32.9374	32.9152	112.9720	1
2	33.4773	33.4548	114.8240	2
3	34.0173	33.9944	116.6760	3
4	34.5572	34.5340	118.5280	4
5	35.0972	35.0736	120.3800	5
6	35.6371	35.6132	122.2320	6
7	36.1771	36.1527	124.0840	7
8	36.7171	36.6923	125.9360	8
9	37.2570	37.2319	127.7880	9
70	37.7970	37.7715	129.6400	70
1	38.3369	38.3111	131.4920	1
2	38.8769	38.8507	133.3440	2
3	39.4168	39.3903	135.1960	3
4	39.9568	39.9299	137.0480	4
5	40.4968	40.4695	138.9000	5
6	41.0367	41.0091	140.7520	6
7	41.5767	41.5487	142.6040	7
8	42.1166	42.0883	144.4560	8
9	42.6566	42.6279	146.3080	9
80	43.1965	43.1675	148.1600	80
1	43.7365	43.7070	150.0120	1
2	44.2765	44.2466	151.8640	2
3	44.8164	44.7862	153.7160	3
4	45.3564	45.3258	155.5680	4
5	45.8963	45.8654	157.4200	5
6	46.4363	46.4050	159.2720	6
7	46.9762	46.9446	161.1240	7
8	47.5162	47.4842	162.9760	8
9	48.0562	48.0238	164.8280	9
90	48.5961	48.5634	166.6800	90
1	49.1361	49.1030	168.5320	1
2	49.6760	49.6426	170.3840	2
3	50.2160	50.1822	172.2360	3
4	50.7559	50.7218	174.0880	4
5	51.2959	51.2614	175.9400	5
6	51.8359	51.8009	177.7920	6
7	52.3758	52.3405	179.6440	7
8	52.9158	52.8801	181.4960	8
9	53.4557	53.4197	183.3480	9
100	53.9957	53.9593	185.2000	100

	A	B	C	
50	49.9663	92.6624	50.0337	50
1	50.9657	94.5156	51.0344	1
2	51.9650	96.3689	52.0350	2
3	52.9643	98.2221	53.0357	3
4	53.9636	100.0754	54.0364	4
5	54.9630	101.9286	55.0371	5
6	55.9623	103.7819	56.0377	6
7	56.9616	105.6351	57.0384	7
8	57.9609	107.4884	58.0391	8
9	58.9603	109.3416	59.0398	9
60	59.9596	111.1949	60.0404	60
1	60.9589	113.0481	61.0411	1
2	61.9582	114.9014	62.0418	2
3	62.9576	116.7546	63.0425	3
4	63.9569	118.6079	64.0431	4
5	64.9562	120.4611	65.0438	5
6	65.9556	122.3144	66.0445	6
7	66.9549	124.1676	67.0452	7
8	67.9542	126.0209	68.0458	8
9	68.9535	127.8741	69.0465	9
70	69.9529	129.7274	70.0472	70
1	70.9522	131.5806	71.0478	1
2	71.9515	133.4339	72.0485	2
3	72.9508	135.2871	73.0492	3
4	73.9502	137.1404	74.0499	4
5	74.9495	138.9936	75.0505	5
6	75.9488	140.8468	76.0512	6
7	76.9481	142.7001	77.0519	7
8	77.9475	144.5533	78.0526	8
9	78.9468	146.4066	79.0532	9
80	79.9461	148.2598	80.0539	80
1	80.9455	150.1131	81.0546	1
2	81.9448	151.9663	82.0553	2
3	82.9441	153.8196	83.0559	3
4	83.9434	155.6728	84.0566	4
5	84.9428	157.5261	85.0573	5
6	85.9421	159.3793	86.0580	6
7	86.9414	161.2326	87.0586	7
8	87.9407	163.0858	88.0593	8
9	88.9401	164.9391	89.0600	9
90	89.9394	166.7923	90.0607	90
1	90.9387	168.6456	91.0613	1
2	91.9380	170.4988	92.0620	2
3	92.9374	172.3521	93.0627	3
4	93.9367	174.2053	94.0633	4
5	94.9360	176.0586	95.0640	5
6	95.9354	177.9118	96.0647	6
7	96.9347	179.7651	97.0654	7
8	97.9340	181.6183	98.0660	8
9	98.9333	183.4716	99.0667	9
100	99.9337	185.3248	100.0674	100

IX. LENGTH—MILLIMETRES TO INCHES

[From 0.00 to 25.40 millimetres by 0.01 millimetre. 1 millimetre = 0.03937 inch.]

Milli-metres	Hundredths of millimetres				
	0.00	**0.01**	**0.02**	**0.03**	**0.04**
	Inches	Inches	Inches	Inches	Inches
0.00	0.000000	0.000394	0.000787	0.001181	0.001575
0.10	.003937	.004331	.004724	.005118	.005512
0.20	.007874	.008268	.008661	.009055	.009449
0.30	.011811	.012205	.012598	.012992	.013386
0.40	.015748	.016142	.016535	.016929	.017323
0.50	0.019685	0.020079	0.020472	0.020866	0.021260
0.60	.023622	.024016	.024409	.024803	.025197
0.70	.027559	.027953	.028346	.028740	.029134
0.80	.031496	.031890	.032283	.032677	.033071
0.90	.035433	.035827	.036220	.036614	.037008
1.00	0.03937	0.03976	0.04016	0.04055	0.04094
1.10	.04331	.04370	.04409	.04449	.04488
1.20	.04724	.04764	.04803	.04843	.04882
1.30	.05118	.05157	.05197	.05236	.05276
1.40	.05512	.05551	.05591	.05630	.05669
1.50	0.05906	0.05945	0.05984	0.06024	0.06063
1.60	.06299	.06339	.06378	.06417	.06457
1.70	.06693	.06732	.06772	.06811	.06850
1.80	.07087	.07126	.07165	.07205	.07244
1.90	.07480	.07520	.07559	.07598	.07638
2.00	0.07874	0.07913	0.07953	0.07992	0.08031
2.10	.08268	.08307	.08346	.08386	.08425
2.20	.08661	.08701	.08740	.08780	.08819
2.30	.09055	.09094	.09134	.09173	.09213
2.40	.09449	.09488	.09528	.09567	.09606
2.50	0.09842	0.09882	0.09921	0.09961	0.10000
2.60	.10236	.10276	.10315	.10354	.10394
2.70	.10630	.10669	.10709	.10748	.10787
2.80	.11024	.11063	.11102	.11142	.11181
2.90	.11417	.11457	.11496	.11535	.11575
3.00	0.11811	0.11850	0.11890	0.11929	0.11968
3.10	.12205	.12244	.12283	.12323	.12362
3.20	.12598	.12638	.12677	.12717	.12756
3.30	.12992	.13031	.13071	.13110	.13150
3.40	.13386	.13425	.13465	.13504	.13543

IX. LENGTH—MILLIMETRES TO INCHES—continued

[From 0.00 to 25.40 millimetres by 0.01 millimetre. 1 millimetre = 0.03937 inch.]

Milli-metres	Hundredths of millimetres				
	0.05	**0.06**	**0.07**	**0.08**	**0.09**
	Inches	Inches	Inches	Inches	Inches
0.00	0. 001968	0. 002362	0. 002756	0. 003150	0. 003543
0. 10	. 005906	. 006299	. 006693	. 007087	. 007480
0. 20	. 009842	. 010236	. 010630	. 011024	. 011417
0. 30	. 013780	. 014173	. 014567	. 014961	. 015354
0. 40	. 017716	. 018110	. 018504	. 018898	. 019291
0.50	0. 021654	0. 022047	0. 022441	0. 022835	0. 023228
0. 60	. 025590	. 025984	. 026378	. 026772	. 027165
0. 70	. 029528	. 029921	. 030315	. 030709	. 031102
0. 80	. 033464	. 033858	. 034252	. 034646	. 035039
0. 90	. 037402	. 037795	. 038189	. 038583	. 038976
1.00	0. 04134	0. 04173	0. 04213	0. 04252	0. 04291
1. 10	. 04528	. 04567	. 04606	. 04646	. 04685
1. 20	. 04921	. 04961	. 05000	. 05039	. 05079
1. 30	. 05315	. 05354	. 05394	. 05433	. 05472
1. 40	. 05709	. 05748	. 05787	. 05827	. 05866
1.50	0. 06102	0. 06142	0. 06181	0. 06220	0. 06260
1. 60	. 06496	. 06535	. 06575	. 06614	. 06654
1. 70	. 06890	. 06929	. 06968	. 07008	. 07047
1. 80	. 07283	. 07323	. 07362	. 07402	. 07441
1. 90	. 07677	. 07717	. 07756	. 07795	. 07835
2.00	0. 08071	0. 08110	0. 08150	0. 08189	0. 08228
2. 10	. 08465	. 08504	. 08543	. 08583	. 08622
2. 20	. 08858	. 08898	. 08937	. 08976	. 09016
2. 30	. 09252	. 09291	. 09331	. 09370	. 09409
2. 40	. 09646	. 09685	. 09724	. 09764	. 09803
2.50	0. 10039	0. 10079	0. 10118	0. 10157	0. 10197
2. 60	. 10433	. 10472	. 10512	. 10551	. 10591
2. 70	. 10827	. 10866	. 10905	. 10945	. 10984
2. 80	. 11220	. 11260	. 11299	. 11339	. 11378
2. 90	. 11614	. 11654	. 11693	. 11732	. 11772
3.00	0. 12008	0. 12047	0. 12087	0. 12126	0. 12165
3. 10	. 12402	. 12441	. 12480	. 12520	. 12559
3. 20	. 12795	. 12835	. 12874	. 12913	. 12953
3. 30	. 13189	. 13228	. 13268	. 13307	. 13346
3. 40	. 13583	. 13622	. 13661	. 13701	. 13740

[From 0.00 to 25.40 millimetres by 0.01 millimetre. 1 millimetre = 0.03937 inch.]

Milli-metres	Hundredths of millimetres				
	0.00	0.01	0.02	0.03	0.04
	Inches	Inches	Inches	Inches	Inches
3.50	0. 13780	0. 13819	0. 13858	0. 13898	0. 13937
3. 60	. 14173	. 14213	. 14252	. 14291	. 14331
3. 70	. 14567	. 14606	. 14646	. 14685	. 14724
3. 80	. 14961	. 15000	. 15039	. 15079	. 15118
3. 90	. 15354	. 15394	. 15433	. 15472	. 15512
4.00	0. 15748	0. 15787	0. 15827	0. 15866	0. 15905
4. 10	. 16142	. 16181	. 16220	. 16260	. 16299
4. 20	. 16535	. 16575	. 16614	. 16654	. 16693
4. 30	. 16929	. 16968	. 17008	. 17047	. 17087
4. 40	. 17323	. 17362	. 17402	. 17441	. 17480
4.50	0. 17716	0. 17756	0. 17795	0. 17835	0. 17874
4. 60	. 18110	. 18150	. 18189	. 18228	. 18268
4. 70	. 18504	. 18543	. 18583	. 18622	. 18661
4. 80	. 18898	. 18937	. 18976	. 19016	. 19055
4. 90	. 19291	. 19331	. 19370	. 19409	. 19449
5.00	0. 19685	0. 19724	0. 19764	0. 19803	0. 19842
5. 10	. 20079	. 20118	. 20157	. 20197	. 20236
5. 20	. 20472	. 20512	. 20551	. 20591	. 20630
5. 30	. 20866	. 20905	. 20945	. 20984	. 21024
5. 40	. 21260	. 21299	. 21339	. 21378	. 21417
5.50	0. 21654	0. 21693	0. 21732	0. 21772	0. 21811
5. 60	. 22047	. 22087	. 22126	. 22165	. 22205
5. 70	. 22441	. 22480	. 22520	. 22559	. 22598
5. 80	. 22835	. 22874	. 22913	. 22953	. 22992
5. 90	. 23228	. 23268	. 23307	. 23346	. 23386
6.00	0. 23622	0. 23661	0. 23701	0. 23740	0. 23779
6. 10	. 24016	. 24055	. 24094	. 24134	. 24173
6. 20	. 24409	. 24449	. 24488	. 24528	. 24567
6. 30	. 24803	. 24842	. 24882	. 24921	. 24961
6. 40	. 25197	. 25236	. 25276	. 25315	. 25354
6.50	0. 25590	0. 25630	0. 25669	0. 25709	0. 25748
6. 60	. 25984	. 26024	. 26063	. 26102	. 26142
6. 70	. 26378	. 26417	. 26457	. 26496	. 26535
6. 80	. 26772	. 26811	. 26850	. 26890	. 26929
6. 90	. 27165	. 27205	. 27244	. 27283	. 27323

[From 0.00 to 25.40 millimetres by 0.01 millimetre. 1 millimetre = 0.03937 inch.]

Milli- metres	Hundredths of millimetres				
	0.05	0.06	0.07	0.08	0.09
	Inches	Inches	Inches	Inches	Inches
3.50	0. 13976	0. 14016	0. 14055	0. 14094	0. 14134
3. 60	. 14370	. 14409	. 14449	. 14488	. 14528
3. 70	. 14764	. 14803	. 14842	. 14882	. 14921
3. 80	. 15157	. 15197	. 15236	. 15276	. 15315
3. 90	. 15551	. 15591	. 15630	. 15669	. 15709
4.00	0. 15945	0. 15984	0. 16024	0. 16063	0. 16102
4. 10	. 16339	. 16378	. 16417	. 16457	. 16496
4. 20	. 16732	. 16772	. 16811	. 16850	. 16890
4. 30	. 17126	. 17165	. 17205	. 17244	. 17283
4. 40	. 17520	. 17559	. 17598	. 17638	. 17677
4.50	0. 17913	0. 17953	0. 17992	0. 18031	0. 18071
4. 60	. 18307	. 18346	. 18386	. 18425	. 18465
4. 70	. 18701	. 18740	. 18779	. 18819	. 18858
4. 80	. 19094	. 19134	. 19173	. 19213	. 19252
4. 90	. 19488	. 19528	. 19567	. 19606	. 19646
5.00	0. 19882	0. 19921	0. 19961	0. 20000	0. 20039
5. 10	. 20276	. 20315	. 20354	. 20394	. 20433
5. 20	. 20669	. 20709	. 20748	. 20787	. 20827
5. 30	. 21063	. 21102	. 21142	. 21181	. 21220
5. 40	. 21457	. 21496	. 21535	. 21575	. 21614
5.50	0. 21850	0. 21890	0. 21929	0. 21968	0. 22008
5. 60	. 22244	. 22283	. 22323	. 22362	. 22402
5. 70	. 22638	. 22677	. 22716	. 22756	. 22795
5. 80	. 23031	. 23071	. 23110	. 23150	. 23189
5. 90	. 23425	. 23465	. 23504	. 23543	. 23583
6.00	0. 23819	0. 23858	0. 23898	0. 23937	0. 23976
6. 10	. 24213	. 24252	. 24291	. 24331	. 24370
6. 20	. 24606	. 24646	. 24685	. 24724	. 24764
6. 30	. 25000	. 25039	. 25079	. 25118	. 25157
6. 40	. 25394	. 25433	. 25472	. 25512	. 25551
6.50	0. 25787	0. 25827	0. 25866	0. 25905	0. 25945
6. 60	. 26181	. 26220	. 26260	. 26299	. 26339
6. 70	. 26575	. 26614	. 26653	. 26693	. 26732
6. 80	. 26968	. 27008	. 27047	. 27087	. 27126
6. 90	. 27362	. 27402	. 27441	. 27480	. 27520

[From 0.00 to 25.40 millimetres by 0.01 millimetre. 1 millimetre = 0.03937 inch.]

Milli-metres	Hundredths of millimetres				
	0.00	0.01	0.02	0.03	0.04
	Inches	Inches	Inches	Inches	Inches
7.00	0.27559	0.27598	0.27638	0.27677	0.27716
7.10	.27953	.27992	.28031	.28071	.28110
7.20	.28346	.28386	.28425	.28465	.28504
7.30	.28740	.28779	.28819	.28858	.28898
7.40	.29134	.29173	.29213	.29252	.29291
7.50	0.29528	0.29567	0.29606	0.29646	0.29685
7.60	.29921	.29961	.30000	.30039	.30079
7.70	.30315	.30354	.30394	.30433	.30472
7.80	.30709	.30748	.30787	.30827	.30866
7.90	.31102	.31142	.31181	.31220	.31260
8.00	0.31496	0.31535	0.31575	0.31614	0.31653
8.10	.31890	.31929	.31968	.32008	.32047
8.20	.32283	.32323	.32362	.32402	.32441
8.30	.32677	.32716	.32756	.32795	.32835
8.40	.33071	.33110	.33150	.33189	.33228
8.50	0.33464	0.33504	0.33543	0.33583	0.33622
8.60	.33858	.33898	.33937	.33976	.34016
8.70	.34252	.34291	.34331	.34370	.34409
8.80	.34646	.34685	.34724	.34764	.34803
8.90	.35039	.35079	.35118	.35157	.35197
9.00	0.35433	0.35472	0.35512	0.35551	0.35590
9.10	.35827	.35866	.35905	.35945	.35984
9.20	.36220	.36260	.36299	.36339	.36378
9.30	.36614	.36653	.36693	.36732	.36772
9.40	.37008	.37047	.37087	.37126	.37165
9.50	0.37402	0.37441	0.37480	0.37520	0.37559
9.60	.37795	.37835	.37874	.37913	.37953
9.70	.38189	.38228	.38268	.38307	.38346
9.80	.38583	.38622	.38661	.38701	.38740
9.90	.38976	.39016	.39055	.39094	.39134
10.00	0.39370	0.39409	0.39449	0.39488	0.39527
10.10	.39764	.39803	.39842	.39882	.39921
10.20	.40157	.40197	.40236	.40276	.40315
10.30	.40551	.40590	.40630	.40669	.40709
10.40	.40945	.40984	.41024	.41063	.41102

[From 0.00 to 25.40 millimetres by 0.01 millimetre. 1 millimetre = 0.03937 inch.]

Milli- metres	Hundredths of millimetres				
	0.05	0.06	0.07	0.08	0.09
	Inches	Inches	Inches	Inches	Inches
7.00	0.27756	0.27795	0.27835	0.27874	0.27913
7.10	.28150	.28189	.28228	.28268	.28307
7.20	.28543	.28583	.28622	.28661	.28701
7.30	.28937	.28976	.29016	.29055	.29094
7.40	.29331	.29370	.29409	.29449	.29488
7.50	0.29724	0.29764	0.29803	0.29842	0.29882
7.60	.30118	.30157	.30197	.30236	.30276
7.70	.30512	.30551	.30590	.30630	.30669
7.80	.30905	.30945	.30984	.31024	.31063
7.90	.31299	.31339	.31378	.31417	.31457
8.00	0.31693	0.31732	0.31772	0.31811	0.31850
8.10	.32087	.32126	.32165	.32205	.32244
8.20	.32480	.32520	.32559	.32598	.32638
8.30	.32874	.32913	.32953	.32992	.33031
8.40	.33268	.33307	.33346	.33386	.33425
8.50	0.33661	0.33701	0.33740	0.33779	0.33819
8.60	.34055	.34094	.34134	.34173	.34213
8.70	.34449	.34488	.34527	.34567	.34606
8.80	.34842	.34882	.34921	.34961	.35000
8.90	.35236	.35276	.35315	.35354	.35394
9.00	0.35630	0.35669	0.35709	0.35748	0.35787
9.10	.36024	.36063	.36102	.36142	.36181
9.20	.36417	.36457	.36496	.36535	.36575
9.30	.36811	.36850	.36890	.36929	.36968
9.40	.37205	.37244	.37283	.37323	.37362
9.50	0.37598	0.37638	0.37677	0.37716	0.37756
9.60	.37992	.38031	.38071	.38110	.38150
9.70	.38386	.38425	.38464	.38504	.38543
9.80	.38779	.38819	.38858	.38898	.38937
9.90	.39173	.39213	.39252	.39291	.39331
10.00	0.39567	0.39606	0.39646	0.39685	0.39724
10.10	.39961	.40000	.40039	.40079	.40118
10.20	.40354	.40394	.40433	.40472	.40512
10.30	.40748	.40787	.40827	.40866	.40905
10.40	.41142	.41181	.41220	.41260	.41299

[From 0.00 to 25.40 millimetres by 0.01 millimetre. 1 millimetre = 0.03937 inch.]

Milli-metres	Hundredths of millimetres				
	0.00	0.01	0.02	0.03	0.04
	Inches	Inches	Inches	Inches	Inches
10.50	0. 41338	0. 41378	0. 41417	0. 41457	0. 41496
10. 60	. 41732	. 41772	. 41811	. 41850	. 41890
10. 70	. 42126	. 42165	. 42205	. 42244	. 42283
10. 80	. 42520	. 42559	. 42598	. 42638	. 42677
10. 90	. 42913	. 42953	. 42992	. 43031	. 43071
11.00	0. 43307	0. 43346	0. 43386	0. 43425	0. 43464
11. 10	. 43701	. 43740	. 43779	. 43819	. 43858
11. 20	. 44094	. 44134	. 44173	. 44213	. 44252
11. 30	. 44488	. 44527	. 44567	. 44606	. 44646
11. 40	. 44882	. 44921	. 44961	. 45000	. 45039
11.50	0. 45276	0. 45315	0. 45354	0. 45394	0. 45433
11. 60	. 45669	. 45709	. 45748	. 45787	. 45827
11. 70	. 46063	. 46102	. 46142	. 46181	. 46220
11. 80	. 46457	. 46496	. 46535	. 46575	. 46614
11. 90	. 46850	. 46890	. 46929	. 46968	. 47008
12.00	0. 47244	0. 47283	0. 47323	0. 47362	0. 47401
12. 10	. 47638	. 47677	. 47716	. 47756	. 47795
12. 20	. 48031	. 48071	. 48110	. 48150	. 48189
12. 30	. 48425	. 48464	. 48504	. 48543	. 48583
12. 40	. 48819	. 48858	. 48898	. 48937	. 48976
12.50	0. 49212	0. 49252	0. 49291	0. 49331	0. 49370
12. 60	. 49606	. 49646	. 49685	. 49724	. 49764
12. 70	. 50000	. 50039	. 50079	. 50118	. 50157
12. 80	. 50394	. 50433	. 50472	. 50512	. 50551
12. 90	. 50787	. 50827	. 50866	. 50905	. 50945
13.00	0. 51181	0. 51220	0. 51260	0. 51299	0. 51338
13. 10	. 51575	. 51614	. 51653	. 51693	. 51732
13. 20	. 51968	. 52008	. 52047	. 52087	. 52126
13. 30	. 52362	. 52401	. 52441	. 52480	. 52520
13. 40	. 52756	. 52795	. 52835	. 52874	. 52913
13.50	0. 53150	0. 53189	0. 53228	0. 53268	0. 53307
13. 60	. 53543	. 53583	. 53622	. 53661	. 53701
13. 70	. 53937	. 53976	. 54016	. 54055	. 54094
13. 80	. 54331	. 54370	. 54409	. 54449	. 54488
13. 90	. 54724	. 54764	. 54803	. 54842	. 54882

[From 0.00 to 25.40 millimetres by 0.01 millimetre. 1 millimetre = 0.03937 inch.]

Milli-metres	Hundredths of millimetres				
	0.05	0.06	0.07	0.08	0.09
	Inches	Inches	Inches	Inches	Inches
10.50	0.41535	0.41575	0.41614	0.41653	0.41693
10.60	.41929	.41968	.42008	.42047	.42087
10.70	.42323	.42362	.42401	.42441	.42480
10.80	.42716	.42756	.42795	.42835	.42874
10.90	.43110	.43150	.43189	.43228	.43268
11.00	0.43504	0.43543	0.43583	0.43622	0.43661
11.10	.43898	.43937	.43976	.44016	.44055
11.20	.44291	.44331	.44370	.44409	.44449
11.30	.44685	.44724	.44764	.44803	.44842
11.40	.45079	.45118	.45157	.45197	.45236
11.50	0.45472	0.45512	0.45551	0.45590	0.45630
11.60	.45866	.45905	.45945	.45984	.46024
11.70	.46260	.46299	.46338	.46378	.46417
11.80	.46653	.46693	.46732	.46772	.46811
11.90	.47047	.47087	.47126	.47165	.47205
12.00	0.47441	0.47480	0.47520	0.47559	0.47598
12.10	.47835	.47874	.47913	.47953	.47992
12.20	.48228	.48268	.48307	.48346	.48386
12.30	.48622	.48661	.48701	.48740	.48779
12.40	.49016	.49055	.49094	.49134	.49173
12.50	0.49409	0.49449	0.49488	0.49527	0.49567
12.60	.49803	.49842	.49882	.49921	.49961
12.70	.50197	.50236	.50275	.50315	.50354
12.80	.50590	.50630	.50669	.50709	.50748
12.90	.50984	.51024	.51063	.51102	.51142
13.00	0.51378	0.51417	0.51457	0.51496	0.51535
13.10	.51772	.51811	.51850	.51890	.51929
13.20	.52165	.52205	.52244	.52283	.52323
13.30	.52559	.52598	.52638	.52677	.52716
13.40	.52953	.52992	.53031	.53071	.53110
13.50	0.53346	0.53386	0.53425	0.53464	0.53504
13.60	.53740	.53779	.53819	.53858	.53898
13.70	.54134	.54173	.54212	.54252	.54291
13.80	.54527	.54567	.54606	.54646	.54685
13.90	.54921	.54961	.55000	.55039	.55079

[From 0.00 to 25.40 millimetres by 0.01 millimetre. 1 millimetre = 0.03937 inch.]

Milli-metres	Hundredths of millimetres				
	0.00	0.01	0.02	0.03	0.04
	Inches	Inches	Inches	Inches	Inches
14.00	0. 55118	0. 55157	0. 55197	0. 55236	0. 55275
14. 10	.. 55512	. 55551	. 55590	. 55630	. 55669
14. 20	. 55905	. 55945	. 55984	. 56024	. 56063
14. 30	. 56299	. 56338	. 56378	. 56417	. 56457
14. 40	. 56693	. 56732	. 56772	. 56811	. 56850
14.50	0. 57086	0. 57126	0. 57165	0. 57205	0. 57244
14. 60	. 57480	. 57520	. 57559	. 57598	. 57638
14. 70	. 57874	. 57913	. 57953	. 57992	. 58031
14. 80	. 58268	. 58307	. 58346	. 58386	. 58425
14. 90	. 58661	. 58701	. 58740	. 58779	. 58819
15.00	0. 59055	0. 59094	0. 59134	0. 59173	0. 59212
15. 10	. 59449	. 59488	. 59527	. 59567	. 59606
15. 20	. 59842	. 59882	. 59921	. 59961	. 60000
15. 30	. 60236	. 60275	. 60315	. 60354	. 60394
15. 40	. 60630	. 60669	. 60709	. 60748	. 60787
15.50	0. 61024	0. 61063	0. 61102	0. 61142	0. 61181
15. 60	. 61417	. 61457	. 61496	. 61535	. 61575
15. 70	. 61811	. 61850	. 61890	. 61929	. 61968
15. 80	. 62205	. 62244	. 62283	. 62323	. 62362
15. 90	. 62598	. 62638	. 62677	. 62716	. 62756
16.00	0. 62992	0. 63031	0. 63071	0. 63110	0. 63149
16. 10	. 63386	. 63425	. 63464	. 63504	. 63543
16. 20	. 63779	. 63819	. 63858	. 63898	. 63937
16. 30	. 64173	. 64212	. 64252	. 64291	. 64331
16. 40	. 64567	. 64606	. 64646	. 64685	. 64724
16.50	0. 64960	0. 65000	0. 65039	0. 65079	0. 65118
16. 60	. 65354	. 65394	. 65433	. 65472	. 65512
16. 70	. 65748	. 65787	. 65827	. 65866	. 65905
16. 80	. 66142	. 66181	. 66220	. 66260	. 66299
16. 90	. 66535	. 66575	. 66614	. 66653	. 66693
17.00	0. 66929	0. 66968	0. 67008	0. 67047	0. 67086
17. 10	. 67323	. 67362	. 67401	. 67441	. 67480
17. 20	. 67716	. 67756	. 67795	. 67835	. 67874
17. 30	. 68110	. 68149	. 68189	. 68228	. 68268
17. 40	. 68504	. 68543	. 68583	. 68622	. 68661

[From 0.00 to 25.40 millimetres by 0.01 millimetre. 1 millimetre = 0.03937 inch.]

Milli-metres	Hundredths of millimetres				
	0.05	0.06	0.07	0.08	0.09
	Inches	Inches	Inches	Inches	Inches
14.00	0. 55315	0. 55354	0. 55394	0. 55433	0. 55472
14. 10	. 55709	. 55748	. 55787	. 55827	. 55866
14. 20	. 56102	. 56142	. 56181	. 56220	. 56260
14. 30	. 56496	. 56535	. 56575	. 56614	. 56653
14. 40	. 56890	. 56929	. 56968	. 57008	. 57047
14.50	0. 57283	0. 57323	0. 57362	0. 57401	. 57441
14. 60	. 57677	. 57716	. 57756	. 57795	. 57835
14. 70	. 58071	. 58110	. 58149	. 58189	. 58228
14. 80	. 58464	. 58504	. 58543	. 58583	. 58622
14. 90	. 58858	. 58898	. 58937	. 58976	. 59016
15.00	0. 59252	0. 59291	0. 59331	0. 59370	0. 59409
15. 10	. 59646	. 59685	. 59724	. 59764	. 59803
15. 20	. 60039	. 60079	. 60118	. 60157	. 60197
15. 30	. 60433	. 60472	. 60512	. 60551	. 60590
15. 40	. 60827	. 60866	. 60905	. 60945	. 60984
15.50	0. 61220	0. 61260	0. 61299	0. 61338	0. 61378
15. 60	. 61614	. 61653	. 61693	. 61732	. 61772
15. 70	. 62008	. 62047	. 62086	. 62126	. 62165
15. 80	. 62401	. 62441	. 62480	. 62520	. 62559
15. 90	. 62795	. 62835	. 62874	. 62913	. 62953
16.00	0. 63189	0. 63228	0. 63268	0. 63307	0. 63346
16. 10	. 63583	. 63622	. 63661	. 63701	. 63740
16. 20	. 63976	. 64016	. 64055	. 64094	. 64134
16. 30	. 64370	. 64409	. 64449	. 64488	. 64527
16. 40	. 64764	. 64803	. 64842	. 64882	. 64921
16.50	0. 65157	0. 65197	0. 65236	0. 65275	0. 65315
16. 60	. 65551	. 65590	. 65630	. 65669	. 65709
16. 70	. 65945	. 65984	. 66023	. 66063	. 66102
16. 80	. 66338	. 66378	. 66417	. 66457	. 66496
16. 90	. 66732	. 66772	. 66811	. 66850	. 66890
17.00	0. 67126	0. 67165	0. 67205	0. 67244	0. 67283
17. 10	. 67520	. 67559	. 67598	. 67638	. 67677
17. 20	. 67913	. 67953	. 67992	. 68031	. 68071
17. 30	. 68307	. 68346	. 68386	. 68425	. 68464
17. 40	. 68701	. 68740	. 68779	. 68819	. 68858

[From 0.00 to 25.40 millimetres by 0.01 millimetre. 1 millimetre = 0.03937 inch.]

Milli-metres	Hundredths of millimetres				
	0.00	0.01	0.02	0.03	0.01
	Inches	Inches	Inches	Inches	Inches
17.50	0.68898	0.68937	0.68976	0.69016	0.69055
17.60	.69291	.69331	.69370	.69409	.69449
17.70	.69685	.69724	.69764	.69803	.69842
17.80	.70079	.70118	.70157	.70197	.70236
17.90	.70472	.70512	.70551	.70590	.70630
18.00	0.70866	0.70905	0.70945	0.70984	0.71023
18.10	.71260	.71299	.71338	.71378	.71417
18.20	.71653	.71693	.71732	.71772	.71811
18.30	.72047	.72086	.72126	.72165	.72205
18.40	.72441	.72480	.72520	.72559	.72598
18.50	0.72834	0.72874	0.72913	0.72953	0.72992
18.60	.73228	.73268	.73307	.73346	.73386
18.70	.73622	.73661	.73701	.73740	.73779
18.80	.74016	.74055	.74094	.74134	.74173
18.90	.74409	.74449	.74488	.74527	.74567
19.00	0.74803	0.74842	0.74882	0.74921	0.74960
19.10	.75197	.75236	.75275	.75315	.75354
19.20	.75590	.75630	.75669	.75709	.75748
19.30	.75984	.76023	.76063	.76102	.76142
19.40	.76378	.76417	.76457	.76496	.76535
19.50	0.76772	0.76811	0.76850	0.76890	0.76929
19.60	.77165	.77205	.77244	.77283	.77323
19.70	.77559	.77598	.77638	.77677	.77716
19.80	.77953	.77992	.78031	.78071	.78110
19.90	.78346	.78386	.78425	.78464	.78504
20.00	0.78740	0.78779	0.78819	0.78858	0.78897
20.10	.79134	.79173	.79212	.79252	.79291
20.20	.79527	.79567	.79606	.79646	.79685
20.30	.79921	.79960	.80000	.80039	.80079
20.40	.80315	.80354	.80394	.80433	.80472
20.50	0.80708	0.80748	0.80787	0.80827	0.80866
20.60	.81102	.81142	.81181	.81220	.81260
20.70	.81496	.81535	.81575	.81614	.81653
20.80	.81890	.81929	.81968	.82008	.82047
20.90	.82283	.82323	.82362	.82401	.82441

[From 0.00 to 25.40 millimetres by 0.01 millimetre. 1 millimetre = 0.03937 inch.]

Milli-metres	Hundredths of millimetres				
	0.05	0.06	0.07	0.08	0.09
	Inches	Inches	Inches	Inches	Inches
17.50	0.69094	0.69134	0.69173	0.69212	0.69252
17.60	.69488	.69527	.69567	.69606	.69646
17.70	.69882	.69921	.69960	.70000	.70039
17.80	.70275	.70315	.70354	.70394	.70433
17.90	.70669	.70709	.70748	.70787	.70827
18.00	0.71063	0.71102	0.71142	0.71181	0.71220
18.10	.71457	.71496	.71535	.71575	.71614
18.20	.71850	.71890	.71929	.71968	.72008
18.30	.72244	.72283	.72323	.72362	.72401
18.40	.72638	.72677	.72716	.72756	.72795
18.50	0.73031	0.73071	0.73110	0.73149	0.73189
18.60	.73425	.73464	.73504	.73543	.73583
18.70	.73819	.73858	.73897	.73937	.73976
18.80	.74212	.74252	.74291	.74331	.74370
18.90	.74606	.74646	.74685	.74724	.74764
19.00	0.75000	0.75039	0.75079	0.75118	0.75157
19.10	.75394	.75433	.75472	.75512	.75551
19.20	.75787	.75827	.75866	.75905	.75945
19.30	.76181	.76220	.76260	.76299	.76338
19.40	.76575	.76614	.76653	.76693	.76732
19.50	0.76968	0.77008	0.77047	0.77086	0.77126
19.60	.77362	.77401	.77441	.77480	.77520
19.70	.77756	.77795	.77834	.77874	.77913
19.80	.78149	.78189	.78228	.78268	.78307
19.90	.78543	.78583	.78622	.78661	.78701
20.00	0.78937	0.78976	0.79016	0.79055	0.79094
20.10	.79331	.79370	.79409	.79449	.79488
20.20	.79724	.79764	.79803	.79842	.79882
20.30	.80118	.80157	.80197	.80236	.30275
20.40	.80512	.80551	.80590	.80630	.80669
20.50	0.80905	0.80945	0.80984	0.81023	0.81063
20.60	.81299	.81338	.81378	.81417	.81457
20.70	.81693	.81732	.81771	.81811	.81850
20.80	.82086	.82126	.82165	.82205	.82244
20.90	.82480	.82520	.82559	.82598	.82638

[From 0.00 to 25.40 millimetres by 0.01 millimetre. 1 millimetre = 0.03937 inch.]

Milli-metres	Hundredths of millimetres				
	0.00	0.01	0.02	0.03	0.04
	Inches	Inches	Inches	Inches	Inches
21.00	0.82677	0.82716	0.82756	0.82795	0.82834
21.10	.83071	.83110	.83149	.83189	.83228
21.20	.83464	.83504	.83543	.83583	.83622
21.30	.83858	.83897	.83937	.83976	.84016
21.40	.84252	.84291	.84331	.84370	.84409
21.50	0.84646	0.84685	0.84724	0.84764	0.84803
21.60	.85039	.85079	.85118	.85157	.85197
21.70	.85433	.85472	.85512	.85551	.85590
21.80	.85827	.85866	.85905	.85945	.85984
21.90	.86220	.86260	.86299	.86338	.86378
22.00	0.86614	0.86653	0.86693	0.86732	0.86771
22.10	.87008	.87047	.87086	.87126	.87165
22.20	.87401	.87441	.87480	.87520	.87559
22.30	.87795	.87834	.87874	.87913	.87953
22.40	.88189	.88228	.88268	.88307	.88346
22.50	0.88582	0.88622	0.88661	0.88701	0.88740
22.60	.88976	.89016	.89055	.89094	.89134
22.70	.89370	.89409	.89449	.89488	.89527
22.80	.89764	.89803	.89842	.89882	.89921
22.90	.90157	.90197	.90236	.90275	.90315
23.00	0.90551	0.90590	0.90630	0.90669	0.90708
23.10	.90945	.90984	.91023	.91063	.91102
23.20	.91338	.91378	.91417	.91457	.91496
23.30	.91732	.91771	.91811	.91850	.91890
23.40	.92126	.92165	.92205	.92244	.92283
23.50	0.92520	0.92559	0.92598	0.92638	0.92677
23.60	.92913	.92953	.92992	.93031	.93071
23.70	.93307	.93346	.93386	.93425	.93464
23.80	.93701	.93740	.93779	.93819	.93858
23.90	.94094	.94134	.94173	.94212	.94252
24.00	0.94488	0.94527	0.94567	0.94606	0.94645
24.10	.94882	.94921	.94960	.95000	.95039
24.20	.95275	.95315	.95354	.95394	.95433
24.30	.95669	.95708	.95748	.95787	.95827
24.40	.96063	.96102	.96142	.96181	.96220

[From 0.00 to 25.40 millimetres by 0.01 millimetre. 1 millimetre = 0.03937 inch.]

Milli-metres	Hundredths of millimetres				
	0.05	0.06	0.07	0.08	0.09
	Inches	Inches	Inches	Inches	Inches
21.00	0. 82874	0. 82913	0. 82953	0. 82992	0. 83031
21. 10	. 83268	. 83307	. 83346	. 83386	. 83425
21. 20	. 83661	. 83701	. 83740	. 83779	. 83819
21. 30	. 84055	. 84094	. 84134	. 84173	. 84212
21. 40	. 84449	. 84488	. 84527	. 84567	. 84606
21.50	0. 84842	0. 84882	0. 84921	0. 84960	0. 85000
21. 60	. 85236	. 85275	. 85315	. 85354	. 85394
21. 70	. 85630	. 85669	. 85708	. 85748	. 85787
21. 80	. 86023	. 86063	. 86102	. 86142	. 86181
21. 90	. 86417	. 86457	. 86496	. 86535	. 86575
22.00	0. 86811	0. 86850	0. 86890	0. 86929	0. 86968
22. 10	. 87205	. 87244	. 87283	. 87323	. 87362
22. 20	. 87598	. 87638	. 87677	. 87716	. 87756
22. 30	. 87992	. 88031	. 88071	. 88110	. 88149
22. 40	. 88386	. 88425	. 88464	. 88504	. 88543
22.50	0. 88779	0. 88819	0. 88858	0. 88897	0. 88937
22. 60	. 89173	. 89212	. 89252	. 89291	. 89331
22. 70	. 89567	. 89606	. 89645	. 89685	. 89724
22. 80	. 89960	. 90000	. 90039	. 90079	. 90118
22. 90	. 90354	. 90394	. 90433	. 90472	. 90512
23.00	0. 90748	0. 90787	0. 90827	0. 90866	0. 90905
23. 10	. 91142	. 91181	. 91220	. 91260	. 91299
23. 20	. 91535	. 91575	. 91614	. 91653	. 91693
23. 30	. 91929	. 91968	. 92008	. 92047	. 92086
23. 40	. 92323	. 92362	. 92401	. 92441	. 92480
23.50	0. 92716	0. 92756	0. 92795	0. 92834	0. 92874
23. 60	. 93110	. 93149	. 93189	. 93228	. 93268
23. 70	. 93504	. 93543	. 93582	. 93622	. 93661
23. 80	. 93897	. 93937	. 93976	. 94016	. 94055
23. 90	. 94291	. 94331	. 94370	. 94409	. 94449
24.00	0. 94685	0. 94724	0. 94764	0. 94803	0. 94842
24. 10	. 95079	. 95118	. 95157	. 95197	. 95236
24. 20	. 95472	. 95512	. 95551	. 95590	. 95630
24. 30	. 95866	. 95905	. 95945	. 95984	. 96023
24. 40	. 96260	. 96299	. 96338	. 96378	. 96417

[From 0.00 to 25.40 millimetres by 0.01 millimetre. 1 millimetre = 0.03937 inch.]

Milli-metres	Hundredths of millimetres				
	0.00	0.01	0.02	0.03	0.04
	Inches	Inches	Inches	Inches	Inches
24.50	0.96456	0.96496	0.96535	0.96575	0.96614
24.60	.96850	.96890	.96929	.96968	.97008
24.70	.97244	.97283	.97323	.97362	.97401
24.80	.97638	.97677	.97716	.97756	.97795
24.90	.98031	.98071	.98110	.98149	.98189
25.00	0.98425	0.98464	0.98504	0.98543	0.98582
25.10	.98819	.98858	.98897	.98937	.98976
25.20	.99212	.99252	.99291	.99331	.99370
25.30	.99606	.99645	.99685	.99724	.99764
25.40	1.00000	....	....	....	....

```
 1 inch   = 0.02540 metre
 2 inches =  .05080 metre
 3 inches =  .07620 metre
 4 inches = 0.10160 metre
 5 inches =  .12700 metre
 6 inches =  .15240 metre
 7 inches = 0.17780 metre
 8 inches =  .20320 metre
 9 inches =  .22860 metre
10 inches = 0.25400 metre
11 inches =  .27940 metre
12 inches =  .30480 metre
```

[From 0.00 to 25.40 millimetres by 0.01 millimetre. 1 millimetre = 0.03937 inch.]

Milli-metres	Hundredths of millimetres				
	0.05	0.06	0.07	0.08	0.09
	Inches	Inches	Inches	Inches	Inches
24.50	0.96653	0.96693	0.96732	0.96771	0.96811
24.60	.97047	.97086	.97126	.97165	.97205
24.70	.97441	.97480	.97519	.97559	.97598
24.80	.97834	.97874	.97913	.97953	.97992
24.90	.98228	.98268	.98307	.98346	.98386
25.00	0.98622	0.98661	0.98701	0.98740	0.98779
25.10	.99016	.99055	.99094	.99134	.99173
25.20	.99409	.99449	.99488	.99527	.99567
25.30	.99803	.99842	.99882	.99921	.99960
25.40	....	....	....	....	....

The above tables converting millimetres to inches may be used to convert centimetres to inches by moving the decimal point 1 place to the right, and decimeters to inches by moving the decimal point 2 places to the right.

EXAMPLE: 1 millimetre = 0.03937 inches
1 centimetre = 0.3937   "
1 decimetre = 3.937   "
1 metre   = 39.37   "

IX. LENGTH—METRES TO FEET

[Reduction factor: 1 metre = 3.280833333 feet]

Metres	Feet	Metres	Feet	Metres	Feet	Metres	Feet	Metres	Feet
0		50	164.04167	100	328.08333	150	492.12500	200	656.16667
1	3.28083	1	167.32250	1	331.36417	1	495.40583	1	659.44750
2	6.56167	2	170.60333	2	334.64500	2	498.68667	2	662.72833
3	9.84250	3	173.88417	3	337.92583	3	501.96750	3	666.00917
4	13.12333	4	177.16500	4	341.20667	4	505.24833	4	669.29000
5	16.40417	5	180.44583	5	344.48750	5	508.52917	5	672.57083
6	19.68500	6	183.72667	6	347.76833	6	511.81000	6	675.85167
7	22.96583	7	187.00750	7	351.04917	7	515.09083	7	679.13250
8	26.24667	8	190.28833	8	354.33000	8	518.37167	8	682.41333
9	29.52750	9	193.56917	9	357.61083	9	521.65250	9	685.69417
10	32.80833	60	196.85000	110	360.89167	160	524.93333	210	688.97500
1	36.08917	1	200.13083	1	364.17250	1	528.21417	1	692.25583
2	39.37000	2	203.41167	2	367.45333	2	531.49500	2	695.53667
3	42.65083	3	206.69250	3	370.73417	3	534.77583	3	698.81750
4	45.93167	4	209.97333	4	374.01500	4	538.05667	4	702.09833
5	49.21250	5	213.25417	5	377.29583	5	541.33750	5	705.37917
6	52.49333	6	216.53500	6	380.57667	6	544.61833	6	708.66000
7	55.77417	7	219.81583	7	383.85750	7	547.89917	7	711.94083
8	59.05500	8	223.09667	8	387.13833	8	551.18000	8	715.22167
9	62.33583	9	226.37750	9	390.41917	9	554.46083	9	718.50250
20	65.61667	70	229.65833	120	393.70000	170	557.74167	220	721.78333
1	68.89750	1	232.93917	1	396.98083	1	561.02250	1	725.06417
2	72.17833	2	236.22000	2	400.26167	2	564.30333	2	728.34500
3	75.45917	3	239.50083	3	403.54250	3	567.58417	3	731.62583
4	78.74000	4	242.78167	4	406.82333	4	570.86500	4	734.90667
5	82.02083	5	246.06250	5	410.10417	5	574.14583	5	738.18750
6	85.30167	6	249.34333	6	413.38500	6	577.42667	6	741.46833
7	88.58250	7	252.62417	7	416.66583	7	580.70750	7	744.74917
8	91.86333	8	255.90500	8	419.94667	8	583.98833	8	748.03000
9	95.14417	9	259.18583	9	423.22750	9	587.26917	9	751.31083
30	98.42500	80	262.46667	130	426.50833	180	590.55000	230	754.59167
1	101.70583	1	265.74750	1	429.78917	1	593.83083	1	757.87250
2	104.98667	2	269.02833	2	433.07000	2	597.11167	2	761.15333
3	108.26750	3	272.30917	3	436.35083	3	600.39250	3	764.43417
4	111.54833	4	275.59000	4	439.63167	4	603.67333	4	767.71500
5	114.82917	5	278.87083	5	442.91250	5	606.95417	5	770.99583
6	118.11000	6	282.15167	6	446.19333	6	610.23500	6	774.27667
7	121.39083	7	285.43250	7	449.47417	7	613.51583	7	777.55750
8	124.67167	8	288.71333	8	452.75500	8	616.79667	8	780.83833
9	127.95250	9	291.99417	9	456.03583	9	620.07750	9	784.11917
40	131.23333	90	295.27500	140	459.31667	190	623.35833	240	787.40000
1	134.51417	1	298.55583	1	462.59750	1	626.63917	1	790.68083
2	137.79500	2	301.83667	2	465.87833	2	629.92000	2	793.96167
3	141.07583	3	305.11750	3	469.15917	3	633.20083	3	797.24250
4	144.35667	4	308.39833	4	472.44000	4	636.48167	4	800.52333
5	147.61750	5	311.67917	5	475.72083	5	639.76250	5	803.80417
6	150.91833	6	314.96000	6	479.00167	6	643.04333	6	807.08500
7	154.19917	7	318.24083	7	482.28250	7	646.32417	7	810.36583
8	157.48000	8	321.52167	8	485.56333	8	649.60500	8	813.64667
9	160.76083	9	324.80250	9	488.84417	9	652.88583	9	816.92750

IX. LENGTH—METRES TO FEET

[Reduction factor: 1 metre = 3.280833333 feet]

Metres	Feet	Metres	Feet	Metres	Feet	Metres	Feet	Metres	Feet
250	820.20833	300	984.25000	350	1,148.29167	400	1,312.33333	450	1,476.37500
1	823.48917	1	987.53083	1	1,151.57250	1	1,315.61417	1	1,479.65583
2	826.77000	2	990.81167	2	1,154.85333	2	1,318.89500	2	1,482.93667
3	830.05083	3	994.09250	3	1,158.13417	3	1,322.17583	3	1,486.21750
4	833.33167	4	997.37333	4	1,161.41500	4	1,325.45667	4	1,489.49833
5	836.61250	5	1,000.65417	5	1,164.69583	5	1,328.73750	5	1,492.77917
6	839.89333	6	1,003.93500	6	1,167.97667	6	1,332.01833	6	1,496.06000
7	843.17417	7	1,007.21583	7	1,171.25750	7	1,335.29917	7	1,499.34083
8	846.45500	8	1,010.49667	8	1,174.53833	8	1,338.58000	8	1,502.62167
9	849.73583	9	1,013.77750	9	1,177.81917	9	1,341.86083	9	1,505.90250
260	853.01667	310	1,017.05833	360	1,181.10000	410	1,345.14167	460	1,509.18333
1	856.29750	1	1,020.33917	1	1,184.38083	1	1,348.42250	1	1,512.46417
2	859.57833	2	1,023.62000	2	1,187.66167	2	1,351.70333	2	1,515.74500
3	862.85917	3	1,026.90083	3	1,190.94250	3	1,354.98417	3	1,519.02583
4	866.14000	4	1,030.18167	4	1,194.22333	4	1,358.26500	4	1,522.30667
5	869.42083	5	1,033.46250	5	1,197.50417	5	1,361.54583	5	1,525.58750
6	872.70167	6	1,036.74333	6	1,200.78500	6	1,364.82667	6	1,528.86833
7	875.98250	7	1,040.02417	7	1,204.06583	7	1,368.10750	7	1,532.14917
8	879.26333	8	1,043.30500	8	1,207.34667	8	1,371.38833	8	1,535.43000
9	882.54417	9	1,046.58583	9	1,210.62750	9	1,374.66917	9	1,538.71083
270	885.82500	320	1,049.86667	370	1,213.90833	420	1,377.95000	470	1,541.99167
1	889.10583	1	1,053.14750	1	1,217.18917	1	1,381.23083	1	1,545.27250
2	892.38667	2	1,056.42833	2	1,220.47000	2	1,384.51167	2	1,548.55333
3	895.66750	3	1,059.70917	3	1,223.75083	3	1,387.79250	3	1,551.83417
4	898.94833	4	1,062.99000	4	1,227.03167	4	1,391.07333	4	1,555.11500
5	902.22917	5	1,066.27083	5	1,230.31250	5	1,394.35417	5	1,558.39583
6	905.51000	6	1,069.55167	6	1,233.59333	6	1,397.63500	6	1,561.67667
7	908.79083	7	1,072.83250	7	1,236.87417	7	1,400.91583	7	1,564.95750
8	912.07167	8	1,076.11333	8	1,240.15500	8	1,404.19667	8	1,568.23833
9	915.35250	9	1,079.39417	9	1,243.43583	9	1,407.47750	9	1,571.51917
280	918.63333	330	1,082.67500	380	1,246.71667	430	1,410.75833	480	1,574.80000
1	921.91417	1	1,085.95583	1	1,249.99750	1	1,414.03917	1	1,578.08083
2	925.19500	2	1,089.23667	2	1,253.27833	2	1,417.32000	2	1,581.36167
3	928.47583	3	1,092.51750	3	1,256.55917	3	1,420.60083	3	1,584.64250
4	931.75667	4	1,095.79833	4	1,259.84000	4	1,423.88167	4	1,587.92333
5	935.03750	5	1,099.07917	5	1,263.12083	5	1,427.16250	5	1,591.20417
6	938.31833	6	1,102.36000	6	1,266.40167	6	1,430.44333	6	1,594.48500
7	941.59917	7	1,105.64083	7	1,269.68250	7	1,433.72417	7	1,597.76583
8	944.88000	8	1,108.92167	8	1,272.96333	8	1,437.00500	8	1,601.04667
9	948.16083	9	1,112.20250	9	1,276.24417	9	1,440.28583	9	1,604.32750
290	951.44167	340	1,115.48333	390	1,279.52500	440	1,443.56667	490	1,607.60833
1	954.72250	1	1,118.76417	1	1,282.80583	1	1,446.84750	1	1,610.88917
2	958.00333	2	1,122.04500	2	1,286.08667	2	1,450.12833	2	1,614.17000
3	961.28417	3	1,125.32583	3	1,289.36750	3	1,453.40917	3	1,617.45083
4	964.56500	4	1,128.60667	4	1,292.64833	4	1,456.69000	4	1,620.73167
5	967.84583	5	1,131.88750	5	1,295.92917	5	1,459.97083	5	1,624.01250
6	971.12667	6	1,135.16833	6	1,299.21000	6	1,463.25167	6	1,627.29333
7	974.40750	7	1,138.44917	7	1,302.49083	7	1,466.53250	7	1,630.57417
8	977.68833	8	1,141.73000	8	1,305.77167	8	1,469.81333	8	1,633.85500
9	980.96917	9	1,145.01083	9	1,309.05250	9	1,473.09417	9	1,637.13583

IX. LENGTH—METRES TO FEET

[Reduction factor: 1 metre = 3.280833333 feet]

Metres	Feet	Metres	Feet	Metres	Feet	Metres	Feet	Metres	Feet
500	1,640.41667	550	1,804.45833	600	1,968.50000	650	2,132.54167	700	2,296.58333
1	1,643.69750	1	1,807.73917	1	1,971.78083	1	2,135.82250	1	2,299.86417
2	1,646.97833	2	1,811.02000	2	1,975.06167	2	2,139.10333	2	2,303.14500
3	1,650.25917	3	1,814.30083	3	1,978.34250	3	2,142.38417	3	2,306.42583
4	1,653.54000	4	1,817.58167	4	1,981.62333	4	2,145.66500	4	2,309.70667
5	1,656.82083	5	1,820.86250	5	1,984.90417	5	2,148.94583	5	2,312.98750
6	1,660.10167	6	1,824.14333	6	1,988.18500	6	2,152.22667	6	2,316.26833
7	1,663.38250	7	1,827.42417	7	1,991.46583	7	2,155.50750	7	2,319.54917
8	1,666.66333	8	1,830.70500	8	1,994.74667	8	2,158.78833	8	2,322.83000
9	1,669.94417	9	1,833.98583	9	1,998.02750	9	2,162.06917	9	2,326.11083
510	1,673.22500	560	1,837.26667	610	2,001.30833	660	2,165.35000	710	2,329.39167
1	1,676.50583	1	1,840.54750	1	2,004.58917	1	2,168.63083	1	2,332.67250
2	1,679.78667	2	1,843.82833	2	2,007.87000	2	2,171.91167	2	2,335.95333
3	1,683.06750	3	1,847.10917	3	2,011.15083	3	2,175.19250	3	2,339.23417
4	1,686.34833	4	1,850.39000	4	2,014.43167	4	2,178.47333	4	2,342.51500
5	1,689.62917	5	1,853.67083	5	2,017.71250	5	2,181.75417	5	2,345.79583
6	1,692.91000	6	1,856.95167	6	2,020.99333	6	2,185.03500	6	2,349.07667
7	1,696.19083	7	1,860.23250	7	2,024.27417	7	2,188.31583	7	2,352.35750
8	1,699.47167	8	1,863.51333	8	2,027.55500	8	2,191.59667	8	2,355.63833
9	1,702.75250	9	1,866.79417	9	2,030.83583	9	2,194.87750	9	2,358.91917
520	1,706.03333	570	1,870.07500	620	2,034.11667	670	2,198.15833	720	2,362.20000
1	1,709.31417	1	1,873.35583	1	2,037.39750	1	2,201.43917	1	2,365.48083
2	1,712.59500	2	1,876.63667	2	2,040.67833	2	2,204.72000	2	2,368.76167
3	1,715.87583	3	1,879.91750	3	2,043.95917	3	2,208.00083	3	2,372.04250
4	1,719.15667	4	1,883.19833	4	2,047.24000	4	2,211.28167	4	2,375.32333
5	1,722.43750	5	1,886.47917	5	2,050.52083	5	2,214.56250	5	2,378.60417
6	1,725.71833	6	1,889.76000	6	2,053.80167	6	2,217.84333	6	2,381.88500
7	1,728.99917	7	1,893.04083	7	2,057.08250	7	2,221.12417	7	2,385.16583
8	1,732.28000	8	1,896.32167	8	2,060.36333	8	2,224.40500	8	2,388.44667
9	1,735.56083	9	1,899.60250	9	2,063.64417	9	2,227.68583	9	2,391.72750
530	1,738.84167	580	1,902.88333	630	2,066.92500	680	2,230.96667	730	2,395.00833
1	1,742.12250	1	1,906.16417	1	2,070.20583	1	2,234.24750	1	2,398.28917
2	1,745.40333	2	1,909.44500	2	2,073.48667	2	2,237.52833	2	2,401.57000
3	1,748.68417	3	1,912.72583	3	2,076.76750	3	2,240.80917	3	2,404.85083
4	1,751.96500	4	1,916.00667	4	2,080.04833	4	2,244.09000	4	2,408.13167
5	1,755.24583	5	1,919.28750	5	2,083.32917	5	2,247.37083	5	2,411.41250
6	1,758.52667	6	1,922.56833	6	2,086.61000	6	2,250.65167	6	2,414.69333
7	1,761.80750	7	1,925.84917	7	2,089.89083	7	2,253.93250	7	2,417.97417
8	1,765.08833	8	1,929.13000	8	2,093.17167	8	2,257.21333	8	2,421.25500
9	1,768.36917	9	1,932.41083	9	2,096.45250	9	2,260.49417	9	2,424.53583
540	1,771.65000	590	1,935.69167	640	2,099.73333	690	2,263.77500	740	2,427.81667
1	1,774.93083	1	1,938.97250	1	2,103.01417	1	2,267.05583	1	2,431.09750
2	1,778.21167	2	1,942.25333	2	2,106.29500	2	2,270.33667	2	2,434.37833
3	1,781.49250	3	1,945.53417	3	2,109.57583	3	2,273.61750*	3	2,437.65917
4	1,784.77333	4	1,948.81500	4	2,112.85667	4	2,276.89833*	4	2,440.94000
5	1,788.05417	5	1,952.09583	5	2,116.13750	5	2,280.17917	5	2,444.22083
6	1,791.33500	6	1,955.37667	6	2,119.41833	6	2,283.46000	6	2,447.50167
7	1,794.61583	7	1,958.65750	7	2,122.69917	7	2,286.74083	7	2,450.78250
8	1,797.89667	8	1,961.93833	8	2,125.98000	8	2,290.02167	8	2,454.06333
9	1,801.17750	9	1,965.21917	9	2,129.26083	9	2,293.30250	9	2,457.34417

IX. LENGTH—METRES TO FEET

[Reduction factor: 1 metre = 3.280833333 feet]

Metres	Feet	Metres	Feet	Metres	Feet	Metres	Feet	Metres	Feet
750	2,460.62500	800	2,624.66667	850	2,788.70833	900	2,952.75000	950	3,116.79167
1	2,463.90583	1	2,627.94750	1	2,791.98917	1	2,956.03083	1	3,120.07250
2	2,467.18667	2	2,631.22833	2	2,795.27000	2	2,959.31167	2	3,123.35333
3	2,470.46750	3	2,634.50917	3	2,798.55083	3	2,962.59250	3	3,126.63417
4	2,473.74833	4	2,637.79000	4	2,801.83167	4	2,965.87333	4	3,129.91500
5	2,477.02917	5	2,641.07083	5	2,805.11250	5	2,969.15417	5	3,133.19583
6	2,480.31000	6	2,644.35167	6	2,808.39333	6	2,972.43500	6	3,136.47667
7	2,483.59083	7	2,647.63250	7	2,811.67417	7	2,975.71583	7	3,139.75750
8	2,486.87167	8	2,650.91333	8	2,814.95500	8	2,978.99667	8	3,143.03833
9	2,490.15250	9	2,654.19417	9	2,818.23583	9	2,982.27750	9	3,146.31917
760	2,493 43333	810	2,657.47500	860	2,821.51667	910	2,985.55833	960	3,149.60000
1	2,496.71417	1	2,660.75583	1	2,824.79750	1	2,988.83917	1	3,152.88083
2	2,499.99500	2	2,664.03667	2	2,828.07833	2	2,992.12000	2	3,156.16167
3	2,503.27583	3	2,667.31750	3	2,831.35917	3	2,995.40083	3	3,159.44250
4	2,506.55667	4	2,670.59833	4	2,834.64000	4	2,998.68167	4	3,162.72333
5	2,509.83750	5	2,673.87917	5	2,837.92083	5	3,001.96250	5	3,166.00417
6	2,513.11833	6	2,677.16000	6	2,841.20167	6	3,005.24333	6	3,169.28500
7	2,516.39917	7	2,680.44083	7	2,844.48250	7	3,008.52417	7	3,172.56583
8	2,519.68000	8	2,683.72167	8	2,847.76333	8	3,011.80500	8	3,175.84667
9	2,522.96083	9	2,687.00250	9	2,851.04417	9	3,015.08583	9	3,179.12750
770	2,526.24167	820	2,690.28333	870	2,854.32500	920	3,018.36667	970	3,182.40833
1	2,529.52250	1	2,693.56417	1	2,857.60583	1	3,021.64750	1	3,185.68917
2	2,532.80333	2	2,696.84500	2	2,860.88667	2	3,024.92833	2	3,188.97000
3	2,536.08417	3	2,700.12583	3	2,864.16750	3	3,028.20917	3	3,192.25083
4	2,539.36500	4	2,703.40667	4	2,867.44833	4	3,031.49000	4	3,195.53167
5	2,542.64583	5	2,706.68750	5	2,870.72917	5	3,034.77083	5	3,198.81250
6	2,545.92667	6	2,709.96833	6	2,874.01000	6	3,038.05167	6	3,202.09333
7	2,549.20750	7	2,713.24917	7	2,877.29083	7	3,041.33250	7	3,205.37417
8	2,552.48833	8	2,716.53000	8	2,880.57167	8	3,044.61333	8	3,208.65500
9	2,555.76917	9	2,719.81083	9	2,883.85250	9	3,047.89417	9	3,211.93583
780	2,559.05000	830	2,723.09167	880	2,887.13333	930	3,051.17500	980	3,215.21667
1	2,562.33083	1	2,726.37250	1	2,890.41417	1	3,054.45583	1	3,218.49750
2	2,565.61167	2	2,729.65333	2	2,893.69500	2	3,057.73667	2	3,221.77833
3	2,568.89250	3	2,732.93417	3	2,896.97583	3	3,061.01750	3	3,225.05917
4	2,572.17333	4	2,736.21500	4	2,900.25667	4	3,064.29833	4	3,228.34000
5	2,575.45417	5	2,739.49583	5	2,903.53750	5	3,067.57917	5	3,231.62083
6	2,578.73500	6	2,742.77667	6	2,906.81833	6	3,070.86000	6	3,234.90167
7	2,582.01583	7	2,746.05750	7	2,910.09917	7	3,074.14083	7	3,238.18250
8	2,585.29667	8	2,749.33833	8	2,913.38000	8	3,077.42167	8	3,241.46333
9	2,588.57750	9	2,752.61917	9	2,916.66083	9	3,080.70250	9	3,244.74417
790	2,591.85833	840	2,755.90000	890	2,919.94167	940	3,083.98333	990	3,248.02500
1	2,595.13917	1	2,759.18083	1	2,923.22250	1	3,087.26417	1	3,251.30583
2	2,598.42000	2	2,762.46167	2	2,926.50333	2	3,090.54500	2	3,254.58667
3	2,601.70083	3	2,765.74250	3	2,929.78417	3	3,093.82583	3	3,257.86750
4	2,604.98167	4	2,769.02333	4	2,933.06500	4	3,097.10667	4	3,261.14833
5	2,608.26250	5	2,772.30417	5	2,936.34583	5	3,100.38750	5	3,264.42917
6	2,611.54333	6	2,775.58500	6	2,939.62667	6	3,103.66833	6	3,267.71000
7	2,614.82417	7	2,778.86583	7	2,942.90750	7	3,106.94917	7	3,270.99083
8	2,618.10500	8	2,782.14667	8	2,946.18833	8	3,110.23000	8	3,274.27167
9	2,621.38583	9	2,785.42750	9	2,949.46917	9	3,113.51083	9	3,277.55250

IX. LENGTH—FEET TO METRES

[Reduction factor: 1 foot = 0.3048006096 metre]

Feet	Metres	Feet	Metres	Feet	Metres	Feet	Metres	Feet	Metres
0		50	15.24003	100	30.48006	150	45.72009	200	60.96012
1	0.30480	1	15.54483	1	30.78486	1	46.02489	1	61.26492
2	.60960	2	15.84963	2	31.08966	2	46.32969	2	61.56972
3	.91440	3	16.15443	3	31.39446	3	46.63449	3	61.87452
4	1.21920	4	16.45923	4	31.69926	4	46.93929	4	62.17932
5	1.52400	5	16.76403	5	32.00406	5	47.24409	5	62.48412
6	1.82880	6	17.06883	6	32.30886	6	47.54890	6	62.78893
7	2.13360	7	17.37363	7	32.61367	7	47.85370	7	63.09373
8	2.43840	8	17.67844	8	32.91847	8	48.15850	8	63.39853
9	2.74321	9	17.98324	9	33.22327	9	48.46330	9	63.70333
10	3.04801	60	18.28804	110	33.52807	160	48.76810	210	64.00813
1	3.35281	1	18.59284	1	33.83287	1	49.07290	1	64.31293
2	3.65761	2	18.89764	2	34.13767	2	49.37770	2	64.61773
3	3.96241	3	19.20244	3	34.44247	3	49.68250	3	64.92253
4	4.26721	4	19.50724	4	34.74727	4	49.98730	4	65.22733
5	4.57201	5	19.81204	5	35.05207	5	50.29210	5	65.53213
6	4.87681	6	20.11684	6	35.35687	6	50.59690	6	65.83693
7	5.18161	7	20.42164	7	35.66167	7	50.90170	7	66.14173
8	5.48641	8	20.72644	8	35.96647	8	51.20650	8	66.44653
9	5.79121	9	21.03124	9	36.27127	9	51.51130	9	66.75133
20	6.09601	70	21.33604	120	36.57607	170	51.81610	220	67.05613
1	6.40081	1	21.64084	1	36.88087	1	52.12090	1	67.36093
2	6.70561	2	21.94564	2	37.18567	2	52.42570	2	67.66574
3	7.01041	3	22.25044	3	37.49047	3	52.73051	3	67.97054
4	7.31521	4	22.55525	4	37.79528	4	53.03531	4	68.27534
5	7.62002	5	22.86005	5	38.10008	5	53.34011	5	68.58014
6	7.92482	6	23.16485	6	38.40488	6	53.64491	6	68.88494
7	8.22962	7	23.46965	7	38.70968	7	53.94971	7	69.18974
8	8.53442	8	23.77445	8	39.01448	8	54.25451	8	69.49454
9	8.83922	9	24.07925	9	39.31928	9	54.55931	9	69.79934
30	9.14402	80	24.38405	130	39.62408	180	54.86411	230	70.10414
1	9.44882	1	24.68885	1	39.92888	1	55.16891	1	70.40894
2	9.75362	2	24.99365	2	40.23368	2	55.47371	2	70.71374
3	10.05842	3	25.29845	3	40.53848	3	55.77851	3	71.01854
4	10.36322	4	25.60325	4	40.84328	4	56.08331	4	71.32334
5	10.66802	5	25.90805	5	41.14808	5	56.38811	5	71.62814
6	10.97282	6	26.21285	6	41.45288	6	56.69291	6	71.93294
7	11.27762	7	26.51765	7	41.75768	7	56.99771	7	72.23774
8	11.58242	8	26.82245	8	42.06248	8	57.30251	8	72.54255
9	11.88722	9	27.12725	9	42.36728	9	57.60732	9	72.84735
40	12.19202	90	27.43205	140	42.67209	190	57.91212	240	73.15215
1	12.49682	1	27.73686	1	42.97689	1	58.21692	1	73.45695
2	12.80163	2	28.04166	2	43.28169	2	58.52172	2	73.76175
3	13.10643	3	28.34646	3	43.58649	3	58.82652	3	74.06655
4	13.41123	4	28.65126	4	43.89129	4	59.13132	4	74.37135
5	13.71603	5	28.95606	5	44.19609	5	59.43612	5	74.67615
6	14.02083	6	29.26086	6	44.50089	6	59.74092	6	74.98095
7	14.32563	7	29.56566	7	44.80569	7	60.04572	7	75.28575
8	14.63043	8	29.87046	8	45.11049	8	60.35052	8	75.59055
9	14.93523	9	30.17526	9	45.41529	9	60.65532	9	75.89535

IX. LENGTH—FEET TO METRES

[Reduction factor: 1 foot = 0.3048006096 metre]

Feet	Metres	Feet	Metres	Feet	Metres	Feet	Metres	Feet	Metres
250	76.20015	**300**	91.44018	**350**	106.68021	**400**	121.92024	**450**	137.16027
1	76.50495	1	91.74498	1	106.98501	1	122.22504	1	137.46507
2	76.80975	2	92.04978	2	107.28981	2	122.52985	2	137.76988
3	77.11455	3	92.35458	3	107.59462	3	122.83465	3	138.07468
4	77.41935	4	92.65939	4	107.89942	4	123.13945	4	138.37948
5	77.72416	5	92.96419	5	108.20422	5	123.44425	5	138.68428
6	78.02896	6	93.26899	6	108.50902	6	123.74905	6	138.98908
7	78.33376	7	93.57379	7	108.81382	7	124.05385	7	139.29388
8	78.63856	8	93.87859	8	109.11862	8	124.35865	8	139.59868
9	78.94336	9	94.18339	9	109.42342	9	124.66345	9	139.90348
260	79.24816	**310**	94.48819	**360**	109.72822	**410**	124.96825	**460**	140.20828
1	79.55296	1	94.79299	1	110.03303	1	125.27305	1	140.51308
2	79.85776	2	95.09779	2	110.33782	2	125.57785	2	140.81788
3	80.16256	3	95.40259	3	110.64262	3	125.88265	3	141.12268
4	80.46736	4	95.70739	4	110.94742	4	126.18745	4	141.42748
5	80.77216	5	96.01219	5	111.25222	5	126.49225	5	141.73228
6	81.07696	6	96.31699	6	111.55702	6	126.79705	6	142.03708
7	81.38176	7	96.62179	7	111.86182	7	127.10185	7	142.34188
8	81.68656	8	96.92659	8	112.16662	8	127.40665	8	142.64669
9	81.99136	9	97.23139	9	112.47142	9	127.71146	9	142.95149
270	82.29616	**320**	97.53620	**370**	112.77623	**420**	128.01626	**470**	143.25629
1	82.60097	1	97.84100	1	113.08103	1	128.32106	1	143.56109
2	82.90577	2	98.14580	2	113.38583	2	128.62586	2	143.86589
3	83.21057	3	98.45060	3	113.69063	3	128.93066	3	144.17069
4	83.51537	4	98.75540	4	113.99543	4	129.23546	4	144.47549
5	83.82017	5	99.06020	5	114.30023	5	129.54026	5	144.78029
6	84.12497	6	99.36500	6	114.60503	6	129.84506	6	145.08509
7	84.42977	7	99.66980	7	114.90983	7	130.14986	7	145.38989
8	84.73457	8	99.97460	8	115.21463	8	130.45466	8	145.69469
9	85.03937	9	100.27940	9	115.51943	9	130.75946	9	145.99949
280	85.34417	**330**	100.58420	**380**	115.82423	**430**	131.06426	**480**	146.30429
1	85.64897	1	100.88900	1	116.12903	1	131.36906	1	146.60909
2	85.95377	2	101.19380	2	116.43383	2	131.67386	2	146.91389
3	86.25857	3	101.49860	3	116.73863	3	131.97866	3	147.21869
4	86.56337	4	101.80340	4	117.04343	4	132.28346	4	147.52350
5	86.86817	5	102.10820	5	117.34823	5	132.58827	5	147.82830
6	87.17297	6	102.41300	6	117.65304	6	132.89307	6	148.13310
7	87.47777	7	102.71781	7	117.95784	7	133.19787	7	148.43790
8	87.78258	8	103.02261	8	118.26264	8	133.50267	8	148.74270
9	88.08738	9	103.32741	9	118.56744	9	133.80747	9	149.04750
290	88.39218	**340**	103.63221	**390**	118.87224	**440**	134.11227	**490**	149.35230
1	88.69698	1	103.93701	1	119.17704	1	134.41707	1	149.65710
2	89.00178	2	104.24181	2	119.48184	2	134.72187	2	149.96190
3	89.30658	3	104.54661	3	119.78664	3	135.02667	3	150.26670
4	89.61138	4	104.85141	4	120.09144	4	135.33147	4	150.57150
5	89.91618	5	105.15621	5	120.39624	5	135.63627	5	150.87630
6	90.22098	6	105.46101	6	120.70104	6	135.94107	6	151.18110
7	90.52578	7	105.76581	7	121.00584	7	136.24587	7	151.48590
8	90.83058	8	106.07061	8	121.31064	8	136.55067	8	151.79070
9	91.13538	9	106.37541	9	121.61544	9	136.85547	9	152.09550

IX. LENGTH—FEET TO METRES

[Reduction factor: 1 foot = 0.3048006096 metre]

Feet	Metres	Feet	Metres	Feet	Metres	Feet	Metres	Feet	Metres
500	152.40030	550	167.64034	600	182.88037	650	198.12040	700	213.36043
1	152.70511	1	167.94514	1	183.18517	1	198.42520	1	213.66523
2	153.00991	2	168.24994	2	183.48997	2	198.73000	2	213.97003
3	153.31471	3	168.55474	3	183.79477	3	199.03480	3	214.27483
4	153.61951	4	168.85954	4	184.09957	4	199.33960	4	214.57963
5	153.92431	5	169.16434	5	184.40437	5	199.64440	5	214.88443
6	154.22911	6	169.46914	6	184.70917	6	199.94920	6	215.18923
7	154.53391	7	169.77394	7	185.01397	7	200.25400	7	215.49403
8	154.83871	8	170.07874	8	185.31877	8	200.55880	8	215.79883
9	155.14351	9	170.38354	9	185.62357	9	200.86360	9	216.10363
510	155.44831	560	170.68834	610	185.92837	660	201.16840	710	216.40843
1	155.75311	1	170.99314	1	186.23317	1	201.47320	1	216.71323
2	156.05791	2	171.29794	2	186.53797	2	201.77800	2	217.01803
3	156.36271	3	171.60274	3	186.84277	3	202.08280	3	217.32283
4	156.66751	4	171.90754	4	187.14757	4	202.38760	4	217.62764
5	156.97231	5	172.21234	5	187.45237	5	202.69241	5	217.93244
6	157.27711	6	172.51715	6	187.75718	6	202.99721	6	218.23724
7	157.58192	7	172.82195	7	188.06198	7	203.30201	7	218.54204
8	157.88672	8	173.12675	8	188.36678	8	203.60681	8	218.84684
9	158.19152	9	173.43155	9	188.67158	9	203.91161	9	219.15164
520	158.49632	570	173.73635	620	188.97638	670	204.21641	720	219.45644
1	158.80112	1	174.04115	1	189.28118	1	204.52121	1	219.76124
2	159.10592	2	174.34595	2	189.58598	2	204.82601	2	220.06604
3	159.41072	3	174.65075	3	189.89078	3	205.13081	3	220.37084
4	159.71552	4	174.95555	4	190.19558	4	205.43561	4	220.67564
5	160.02032	5	175.26035	5	190.50038	5	205.74041	5	220.98044
6	160.32512	6	175.56515	6	190.80518	6	206.04521	6	221.28524
7	160.62992	7	175.86995	7	191.10998	7	206.35001	7	221.59004
8	160.93472	8	176.17475	8	191.41478	8	206.65481	8	221.89484
9	161.23952	9	176.47955	9	191.71958	9	206.95961	9	222.19964
530	161.54432	580	176.78435	630	192.02438	680	207.26441	730	222.50445
1	161.84912	1	177.08915	1	192.32918	1	207.56922	1	222.80925
2	162.15392	2	177.39395	2	192.63399	2	207.87402	2	223.11405
3	162.45872	3	177.69876	3	192.93879	3	208.17882	3	223.41885
4	162.76353	4	178.00356	4	193.24359	4	208.48362	4	223.72365
5	163.06833	5	178.30836	5	193.54839	5	208.78842	5	224.02845
6	163.37313	6	178.61316	6	193.85319	6	209.09322	6	224.33325
7	163.67793	7	178.91796	7	194.15799	7	209.39802	7	224.63805
8	163.98273	8	179.22276	8	194.46279	8	209.70282	8	224.94285
9	164.28753	9	179.52756	9	194.76759	9	210.00762	9	225.24765
540	164.59233	590	179.83236	640	195.07239	690	210.31242	740	225.55245
1	164.89713	1	180.13716	1	195.37719	1	210.61722	1	225.85725
2	165.20193	2	180.44196	2	195.68199	2	210.92202	2	226.16205
3	165.50673	3	180.74676	3	195.98679	3	211.22682	3	226.46685
4	165.81153	4	181.05156	4	196.29159	4	211.53162	4	226.77165
5	166.11633	5	181.35636	5	196.59639	5	211.83642	5	227.07645
6	166.42113	6	181.66116	6	196.90119	6	212.14122	6	227.38125
7	166.72593	7	181.96596	7	197.20599	7	212.44602	7	227.68606
8	167.03073	8	182.27076	8	197.51080	8	212.75083	8	227.99086
9	167.33553	9	182.57557	9	197.81560	9	213.05563	9	228.29566

IX. LENGTH—FEET TO METRES

[Reduction factor: 1 foot = 0.3048006096 metre]

Feet	Metres	Feet	Metres	Feet	Metres	Feet	Metres	Feet	Metres
750	228.60046	800	243.84049	850	259.08052	900	274.32055	950	289.56058
1	228.90526	1	244.14529	1	259.38532	1	274.62535	1	289.86538
2	229.21006	2	244.45009	2	259.69012	2	274.93015	2	290.17018
3	229.51486	3	244.75489	3	259.99492	3	275.23495	3	290.47498
4	229.81966	4	245.05969	4	260.29972	4	275.53975	4	290.77978
5	230.12446	5	245.36449	5	260.60452	5	275.84455	5	291.08458
6	230.42926	6	245.66929	6	260.90932	6	276.14935	6	291.33938
7	230.73406	7	245.97409	7	261.21412	7	276.45415	7	291.69418
8	231.03886	8	246.27889	8	261.51892	8	276.75895	8	291.99898
9	231.34366	9	246.53369	9	261.82372	9	277.06375	9	292.30378
760	231.64846	810	246.88849	860	262.12852	910	277.36855	960	292.60859
1	231.95326	1	247.19329	1	262.43332	1	277.67336	1	292.91339
2	232.25806	2	247.49809	2	262.73813	2	277.97816	2	293.21819
3	232.56287	3	247.80290	3	263.04293	3	278.28296	3	293.52299
4	232.86767	4	248.10770	4	263.34773	4	278.58776	4	293.82779
5	233.17247	5	248.41250	5	263.65253	5	278.89256	5	294.13259
6	233.47727	6	248.71730	6	263.95733	6	279.19736	6	294.43739
7	233.78207	7	249.02210	7	264.26213	7	279.50216	7	294.74219
8	234.08687	8	249.32690	8	264.56693	8	279.80696	8	295.04699
9	234.39167	9	249.63170	9	264.87173	9	280.11176	9	295.35179
770	234.69647	820	249.93650	870	265.17653	920	280.41656	970	295.65659
1	235.00127	1	250.24130	1	265.48133	1	280.72136	1	295.96139
2	235.30607	2	250.54610	2	265.78613	2	281.02616	2	296.26619
3	235.61087	3	250.85090	3	266.09093	3	281.33096	3	296.57099
4	235.91567	4	251.15570	4	266.39573	4	281.63576	4	296.87579
5	236.22047	5	251.46050	5	266.70053	5	281.94056	5	297.18059
6	236.52527	6	251.76530	6	267.00533	6	282.24536	6	297.48539
7	236.83007	7	252.07010	7	267.31013	7	282.55017	7	297.79020
8	237.13487	8	252.37490	8	267.61494	8	282.85497	8	298.09500
9	237.43967	9	252.67971	9	267.91974	9	283.15977	9	298.39980
780	237.74448	830	252.98451	880	268.22454	930	283.46457	980	298.70460
1	238.04928	1	253.28931	1	268.52934	1	283.76937	1	299.00940
2	238.35408	2	253.59411	2	268.83414	2	284.07417	2	299.31420
3	238.65888	3	253.89891	3	269.13894	3	284.37897	3	299.61900
4	238.96368	4	254.20371	4	269.44374	4	284.68377	4	299.92380
5	239.26848	5	254.50851	5	269.74854	5	284.98857	5	300.22860
6	239.57328	6	254.81331	6	270.05334	6	285.29337	6	300.53340
7	239.87808	7	255.11811	7	270.35814	7	285.59817	7	300.83820
8	240.18288	8	255.42291	8	270.66294	8	285.90297	8	301.14300
9	240.48768	9	255.72771	9	270.96774	9	286.20777	9	301.44780
790	240.79248	840	256.03251	890	271.27254	940	286.51257	990	301.75260
1	241.09728	1	256.33731	1	271.57734	1	286.81737	1	302.05740
2	241.40208	2	256.64211	2	271.88214	2	287.12217	2	302.36220
3	241.70688	3	256.94691	3	272.18694	3	287.42697	3	302.66701
4	242.01168	4	257.25171	4	272.49174	4	287.73178	4	302.97181
5	242.31648	5	257.55652	5	272.79655	5	288.03658	5	303.27661
6	242.62129	6	257.86132	6	273.10135	6	288.34138	6	303.58141
7	242.92609	7	258.16612	7	273.40615	7	288.64618	7	303.88621
8	243.23089	8	258.47092	8	273.71095	8	288.95098	8	304.19101
9	243.53569	9	258.77572	9	274.01575	9	289.25578	9	304.49581

IX. LENGTH—KILOMETRES TO MILES

[Reduction factor: 1 kilometre = 0.6213699495 mile]

Kilo-metres	Miles	Kilo-metres	Miles	Kilo-metres	Miles	Kilo-metres	Miles	Kilo-metres	Miles
0		50	31.06850	100	62.13699	150	93.20549	200	124.27399
1	0.62137	1	31.68987	1	62.75836	1	93.82686	1	124.89536
2	1.24274	2	32.31124	2	63.37973	2	94.44823	2	125.51673
3	1.86411	3	32.93261	3	64.00110	3	95.06960	3	126.13810
4	2.48548	4	33.55398	4	64.62247	4	95.69097	4	126.75947
5	3.10685	5	34.17535	5	65.24384	5	96.31234	5	127.38084
6	3.72822	6	34.79672	6	65.86521	6	96.93371	6	128.00221
7	4.34959	7	35.41809	7	66.48658	7	97.55508	7	128.62358
8	4.97096	8	36.03946	8	67.10795	8	98.17645	8	129.24495
9	5.59233	9	36.66083	9	67.72932	9	98.79782	9	129.86632
10	6.21370	60	37.28220	110	68.35069	160	99.41919	210	130.48769
1	6.83507	1	37.90357	1	68.97206	1	100.04056	1	131.10906
2	7.45644	2	38.52494	2	69.59343	2	100.66193	2	131.73043
3	8.07781	3	39.14631	3	70.21480	3	101.28330	3	132.35180
4	8.69918	4	39.76768	4	70.83617	4	101.90467	4	132.97317
5	9.32055	5	40.38905	5	71.45754	5	102.52604	5	133.59454
6	9.94192	6	41.01042	6	72.07891	6	103.14741	6	134.21591
7	10.56329	7	41.63179	7	72.70028	7	103.76878	7	134.83728
8	11.18466	8	42.25316	8	73.32165	8	104.39015	8	135.45865
9	11.80603	9	42.87453	9	73.94302	9	105.01152	9	136.08002
20	12.42740	70	43.49590	120	74.56439	170	105.63289	220	136.70139
1	13.04877	1	44.11727	1	75.18576	1	106.25426	1	137.82276
2	13.67014	2	44.73864	2	75.80713	2	106.87563	2	137.94413
3	14.29151	3	45.36001	3	76.42850	3	107.49700	3	138.56550
4	14.91288	4	45.98138	4	77.04987	4	108.11837	4	139.18687
5	15.53425	5	46.60275	5	77.67124	5	108.73974	5	139.80824
6	16.15562	6	47.22412	6	78.29261	6	109.36111	6	140.42961
7	16.77699	7	47.84549	7	78.91398	7	109.98248	7	141.05098
8	17.39836	8	48.46686	8	79.53535	8	110.60385	8	141.67235
9	18.01973	9	49.08823	9	80.15672	9	111.22522	9	142.29372
30	18.64110	80	49.70960	130	80.77809	180	111.84659	230	142.91509
1	19.26247	1	50.33097	1	81.39946	1	112.46796	1	143.53646
2	19.88384	2	50.95234	2	82.02083	2	113.08933	2	144.15783
3	20.50521	3	51.57371	3	82.64220	3	113.71070	3	144.77920
4	21.12658	4	52.19508	4	83.26357	4	114.33207	4	145.40057
5	21.74795	5	52.81645	5	83.88494	5	114.95344	5	146.02194
6	22.36932	6	53.43782	6	84.50631	6	115.57481	6	146.64331
7	22.99069	7	54.05919	7	85.12768	7	116.19618	7	147.26468
8	23.61206	8	54.68056	8	85.74905	8	116.81755	8	147.88605
9	24.23343	9	55.30193	9	86.37042	9	117.43892	9	148.50742
40	24.85480	90	55.92330	140	86.99179	190	118.06029	240	149.12879
1	25.47617	1	56.54467	1	87.61316	1	118.68166	1	149.75016
2	26.09754	2	57.16604	2	88.23453	2	119.30303	2	150.37153
3	26.71891	3	57.78741	3	88.85590	3	119.92440	3	150.99290
4	27.34028	4	58.40878	4	89.47727	4	120.54577	4	151.61427
5	27.96165	5	59.03015	5	90.09864	5	121.16714	5	152.23564
6	28.58302	6	59.65152	6	90.72001	6	121.78851	6	152.85701
7	29.20439	7	60.27289	7	91.34138	7	122.40988	7	153.47838
8	29.82576	8	60.89426	8	91.96275	8	123.03125	8	154.09975
9	30.44713	9	61.51562	9	92.58412	9	123.65262	9	154.72112

IX. LENGTH—KILOMETRES TO MILES

[Reduction factor: 1 kilometre = 0.6213699495 mile]

Kilo-metres	Miles	Kilo-metres	Miles	Kilo-metres	Miles	Kilo-metres	Miles	Kilo-metres	Miles
250	155.34249	300	186.41098	350	217.47948	400	248.54798	450	279.61648
1	155.96386	1	187.03235	1	218.10085	1	249.16935	1	280.23785
2	156.58523	2	187.65372	2	218.72222	2	249.79072	2	280.85922
3	157.20660	3	188.27509	3	219.34359	3	250.41209	3	281.48059
4	157.82797	4	188.89646	4	219.96496	4	251.03346	4	282.10196
5	158.44934	5	189.51783	5	220.58633	5	251.65483	5	282.72333
6	159.07071	6	190.13920	6	221.20770	6	252.27620	6	283.34470
7	159.69208	7	190.76057	7	221.82907	7	252.89757	7	283.96607
8	160.31345	8	191.38194	8	222.45044	8	253.51894	8	284.58744
9	160.93482	9	192.00331	9	223.07181	9	254.14031	9	285.20881
260	161.55619	310	192.62468	360	223.69318	410	254.76168	460	285.83018
1	162.17756	1	193.24605	1	224.31455	1	255.38305	1	286.45155
2	162.79893	2	193.86742	2	224.93592	2	256.00442	2	287.07292
3	163.42030	3	194.48879	3	225.55729	3	256.62579	3	287.69429
4	164.04167	4	195.11016	4	226.17866	4	257.24716	4	288.31566
5	164.66304	5	195.73153	5	226.80003	5	257.86853	5	288.93703
6	165.28441	6	196.35290	6	227.42140	6	258.48990	6	289.55840
7	165.90578	7	196.97427	7	228.04277	7	259.11127	7	290.17977
8	166.52715	8	197.59564	8	-228.66414	8	259.73264	8	290.80114
9	167.14852	9	198.21701	9	229.28551	9	260.35401	9	291.42251
270	167.76989	320	198.83838	370	229.90688	420	260.97538	470	292.04388
1	168.39126	1	199.45975	1	230.52825	1	261.59675	1	292.66525
2	169.01263	2	200.08112	2	231.14962	2	262.21812	2	293.28662
3	169.63400	3	200.70249	3	231.77099	3	262.83949	3	293.90799
4	170.25537	4	201.32386	4	232.39236	4	263.46086	4	294.52936
5	170.87674	5	201.94523	5	233.01373	5	264.08223	5	295.15073
6	171.49811	6	202.56660	6	233.63510	6	264.70360	6	295.77210
7	172.11948	7	203.18797	7	234.25647	7	265.32497	7	296.39347
8	172.74085	8	203.80934	8	234.87784	8	265.94634	8	297.01484
9	173.36222	9	204.43071	9	235.49921	9	266.56771	9	297.63621
280	173.98359	330	205.05208	380	236.12058	430	267.18908	480	298.25758
1	174.60496	1	205.67345	1	236.74195	1	267.81045	1	298.87895
2	175.22633	2	206.29482	2	237.36332	2	268.43182	2	299.50032
3	175.84770	3	206.91619	3	237.98469	3	269.05319	3	300.12169
4	176.46907	4	207.53756	4	238.60606	4	269.67456	4	300.74306
5	177.09044	5	208.15893	5	239.22743	5	270.29593	5	301.36443
6	177.71181	6	208.78030	6	239.84880	6	270.91730	6	301.98580
7	178.33318	7	209.40167	7	240.47017	7	271.53867	7	302.60717
8	178.95455	8	210.02304	8	241.09154	8	272.16004	8	303.22854
9	179.57592	9	210.64441	9	241.71291	9	272.78141	9	303.84991
290	180.19729	340	211.26578	390	242.33428	440	273.40278	490	304.47128
1	180.81866	1	211.88715	1	242.95565	1	274.02415	1	305.09265
2	181.44003	2	212.50852	2	243.57702	2	274.64552	2	305.71402
3	182.06140	3	213.12989	3	244.19839	3	275.26689	3	306.33539
4	182.68277	4	213.75126	4	244.81976	4	275.88826	4	306.95676
5	183.30414	5	214.37263	5	245.44113	5	276.50963	5	307.57812
6	183.92551	6	214.99400	6	246.06250	6	277.13100	6	308.19949
7	184.54687	7	215.61537	7	246.68387	7	277.75237	7	308.82086
8	185.16824	8	216.23674	8	247.30524	8	278.37374	8	309.44223
9	185.78961	9	216.85811	9	247.92661	9	278.99511	9	310.06360

IX. LENGTH—KILOMETRES TO MILES

[Reduction factor: 1 kilometre = 0.6213699495 mile]

Kilo-metres	Miles	Kilo-metres	Miles	Kilo-metres	Miles	Kilo-metres	Miles	Kilo-metres	Miles
500	310.68497	550	341.75347	600	372.82197	650	403.89047	700	434.95896
1	311.30634	1	342.37484	1	373.44334	1	404.51184	1	435.58033
2	311.92771	2	342.99621	2	374.06471	2	405.13321	2	436.20170
3	312.54908	3	343.61758	3	374.68608	3	405.75458	3	436.82307
4	313.17045	4	344.23895	4	375.30745	4	406.37595	4	437.44444
5	313.79182	5	344.86032	5	375.92882	5	406.99732	5	438.06581
6	314.41319	6	345.48169	6	376.55019	6	407.61869	6	438.68718
7	315.03456	7	346.10306	7	377.17156	7	408.24006	7	439.30855
8	315.65593	8	346.72443	8	377.79293	8	408.86143	8	439.92992
9	316.27730	9	347.34580	9	378.41430	9	409.48280	9	440.55129
510	316.89867	560	347.96717	610	379.03567	660	410.10417	710	441.17266
1	317.52004	1	348.58854	1	379.65704	1	410.72554	1	441.79403
2	318.14141	2	349.20991	2	380.27841	2	411.34691	2	442.41540
3	318.76278	3	349.83128	3	380.89978	3	411.96828	3	443.03677
4	319.38415	4	350.45265	4	381.52115	4	412.58965	4	443.65814
5	320.00552	5	351.07402	5	382.14252	5	413.21102	5	444.27951
6	320.62689	6	351.69539	6	382.76389	6	413.83239	6	444.90088
7	321.24826	7	352.31676	7	383.38526	7	414.45376	7	445.52225
8	321.86963	8	352.93813	8	384.00663	8	415.07513	8	446.14362
9	322.49100	9	353.55950	9	384.62800	9	415.69650	9	446.76499
520	323.11237	570	354.18087	620	385.24937	670	416.31787	720	447.38636
1	323.73374	1	354.80224	1	385.87074	1	416.93924	1	448.00773
2	324.35511	2	355.42361	2	386.49211	2	417.56061	2	448.62910
3	324.97648	3	356.04498	3	387.11348	3	418.18198	3	449.25047
4	325.59785	4	356.66635	4	387.73485	4	418.80335	4	449.87184
5	326.21922	5	357.28772	5	388.35622	5	419.42472	5	450.49321
6	326.84059	6	357.90909	6	388.97759	6	420.04609	6	451.11458
7	327.46196	7	358.53046	7	389.59896	7	420.66746	7	451.73595
8	328.08333	8	359.15183	8	390.22033	8	421.28883	8	452.35732
9	328.70470	9	359.77320	9	390.84170	9	421.91020	9	452.97869
530	329.32607	580	360.39457	630	391.46307	680	422.53157	730	453.60006
1	329.94744	1	361.01594	1	392.08444	1	423.15294	1	454.22143
2	330.56881	2	361.63731	2	392.70581	2	423.77431	2	454.84280
3	331.19018	3	362.25868	3	393.32718	3	424.39568	3	455.46417
4	331.81155	4	362.88005	4	393.94855	4	425.01705	4	456.08554
5	332.43292	5	363.50142	5	394.56992	5	425.63842	5	456.70691
6	333.05429	6	364.12279	6	395.19129	6	426.25979	6	457.32828
7	333.67566	7	364.74416	7	395.81266	7	426.88116	7	457.94965
8	334.29703	8	365.36553	8	396.43403	8	427.50253	8	458.57102
9	334.91840	9	365.98690	9	397.05540	9	428.12390	9	459.19239
540	335.53977	590	366.60827	640	397.67677	690	428.74527	740	459.81376
1	336.16114	1	367.22964	1	398.29814	1	429.36664	1	460.43513
2	336.78251	2	367.85101	2	398.91951	2	429.98801	2	461.05650
3	337.40388	3	368.47238	3	399.54088	3	430.60937	3	461.67787
4	338.02525	4	369.09375	4	400.16225	4	431.23074	4	462.29924
5	338.64662	5	369.71512	5	400.78362	5	431.85211	5	462.92061
6	339.26799	6	370.33649	6	401.40499	6	432.47348	6	463.54198
7	339.88936	7	370.95786	7	402.02636	7	433.09485	7	464.16335
8	340.51073	8	371.57923	8	402.64773	8	433.71622	8	464.78472
9	341.13210	9	372.20060	9	403.26910	9	434.33759	9	465.40609

IX. LENGTH—KILOMETRES TO MILES

[Reduction factor: 1 kilometre = 0.6213699495 mile]

Kilometres	Miles	Kilometres	Miles	Kilometres	Miles	Kilometres	Miles	Kilometres	Miles
750	466.02746	800	497.09596	850	528.16446	900	559.23295	950	590.30145
1	466.64883	1	497.71733	1	528.78583	1	559.85432	1	590.92282
2	467.27020	2	498.33870	2	529.40720	2	560.47569	2	591.54419
3	467.89157	3	498.96007	3	530.02857	3	561.09706	3	592.16556
4	468.51294	4	499.58144	4	530.64994	4	561.71843	4	592.78693
5	469.13431	5	500.20281	5	531.27131	5	562.33980	5	593.40830
6	469.75568	6	500.82418	6	531.89268	6	562.96117	6	594.02967
7	470.37705	7	501.44555	7	532.51405	7	563.58254	7	594.65104
8	470.99842	8	502.06692	8	533.13542	8	564.20391	8	595.27241
9	471.61979	9	502.68829	9	533.75679	9	564.82528	9	595.89378
760	472.24116	810	503.30966	860	534.37816	910	565.44665	960	596.51515
1	472.86253	1	503.93103	1	534.99953	1	566.06802	1	597.13652
2	473.48390	2	504.55240	2	535.62090	2	566.68939	2	597.75789
3	474.10527	3	505.17377	3	536.24227	3	567.31076	3	598.37926
4	474.72664	4	505.79514	4	536.86364	4	567.93213	4	599.00063
5	475.34801	5	506.41651	5	537.48501	5	568.55350	5	599.62200
6	475.96938	6	507.03788	6	538.10638	6	569.17487	6	600.24337
7	476.59075	7	507.65925	7	538.72775	7	569.79624	7	600.86474
8	477.21212	8	508.28062	8	539.34912	8	570.41761	8	601.48611
9	477.83349	9	508.90199	9	539.97049	9	571.03898	9	602.10748
770	478.45486	820	509.52336	870	540.59186	920	571.66035	970	602.72885
1	479.07623	1	510.14473	1	541.21323	1	572.28172	1	603.35022
2	479.69760	2	510.76610	2	541.83460	2	572.90309	2	603.97159
3	480.31897	3	511.38747	3	542.45597	3	573.52446	3	604.59296
4	480.94034	4	512.00884	4	543.07734	4	574.14583	4	605.21433
5	481.56171	5	512.63021	5	543.69871	5	574.76720	5	605.83570
6	482.18308	6	513.25158	6	544.32008	6	575.38857	6	606.45707
7	482.80445	7	513.87295	7	544.94145	7	576.00994	7	607.07844
8	483.42582	8	514.49432	8	545.56282	8	576.63131	8	607.69981
9	484.04719	9	515.11569	9	546.18419	9	577.25268	9	608.32118
780	484.66856	830	515.73706	880	546.80556	930	577.87405	980	608.94255
1	485.28993	1	516.35843	1	547.42693	1	578.49542	1	609.56392
2	485.91130	2	516.97980	2	548.04830	2	579.11679	2	610.18529
3	486.53267	3	517.60117	3	548.66967	3	579.73816	3	610.80666
4	487.15404	4	518.22254	4	549.29104	4	580.35953	4	611.42803
5	487.77541	5	518.84391	5	549.91241	5	580.98090	5	612.04940
6	488.39678	6	519.46528	6	550.53378	6	581.60227	6	612.67077
7	489.01815	7	520.08665	7	551.15515	7	582.22364	7	613.29214
8	489.63952	8	520.70802	8	551.77652	8	582.84501	8	613.91351
9	490.26089	9	521.32939	9	552.39789	9	583.46638	9	614.53488
790	490.88226	840	521.95076	890	553.01926	940	584.08775	990	615.15625
1	491.50363	1	522.57213	1	553.64062	1	584.70912	1	615.77762
2	492.12500	2	523.19350	2	554.26199	2	585.33049	2	616.39899
3	492.74637	3	523.81487	3	554.88336	3	585.95186	3	617.02036
4	493.36774	4	524.43624	4	555.50473	4	586.57323	4	617.64173
5	493.98911	5	525.05761	5	556.12610	5	587.19460	5	618.26310
6	494.61048	6	525.67898	6	556.74747	6	587.81597	6	618.88447
7	495.23185	7	526.30035	7	557.36884	7	588.43734	7	619.50584
8	495.85322	8	526.92172	8	557.99021	8	589.05871	8	620.12721
9	496.47459	9	527.54309	9	558.61158	9	589.68008	9	620.74858

IX. LENGTH—MILES TO KILOMETRES

[Reduction factor: 1 mile = 1.609347219 kilometres]

Miles	Kilo-metres	Miles	Kilo-metres	Miles	Kilo-metres	Miles	Kilo-metres	Miles	Kilo-metres
0		50	80.4674	100	160.9347	150	241.4021	200	321.8694
1	1.6093	1	82.0767	1	162.5441	1	243.0114	1	323.4788
2	3.2187	2	83.6861	2	164.1534	2	244.6208	2	325.0881
3	4.8280	3	85.2954	3	165.7628	3	246.2301	3	326.6975
4	6.4374	4	86.9047	4	167.3721	4	247.8395	4	328.3068
5	8.0467	5	88.5141	5	168.9815	5	249.4488	5	329.9162
6	9.6561	6	90.1234	6	170.5908	6	251.0582	6	331.5255
7	11.2654	7	91.7328	7	172.2002	7	252.6675	7	333.1349
8	12.8748	8	93.3421	8	173.8095	8	254.2769	8	334.7442
9	14.4841	9	94.9515	9	175.4188	9	255.8862	9	336.3536
10	16.0935	60	96.5608	110	177.0282	160	257.4956	210	337.9629
1	17.7028	1	98.1702	1	178.6375	1	259.1049	1	339.5723
2	19.3122	2	99.7795	2	180.2469	2	260.7142	2	341.1816
3	20.9215	3	101.3889	3	181.8562	3	262.3236	3	342.7910
4	22.5309	4	102.9982	4	183.4656	4	263.9329	4	344.4003
5	24.1402	5	104.6076	5	185.0749	5	265.5423	5	346.0097
6	25.7496	6	106.2169	6	186.6843	6	267.1516	6	347.6190
7	27.3589	7	107.8263	7	188.2936	7	268.7610	7	349.2283
8	28.9682	8	109.4356	8	189.9030	8	270.3703	8	350.8377
9	30.5776	9	111.0450	9	191.5123	9	271.9797	9	352.4470
20	32.1869	70	112.6543	120	193.1217	170	273.5890	220	354.0564
1	33.7963	1	114.2637	1	194.7310	1	275.1984	1	355.6657
2	35.4056	2	115.8730	2	196.3404	2	276.8077	2	357.2751
3	37.0150	3	117.4823	3	197.9497	3	278.4171	3	358.8844
4	38.6243	4	119.0917	4	199.5591	4	280.0264	4	360.4938
5	40.2337	5	120.7010	5	201.1684	5	281.6358	5	362.1031
6	41.8430	6	122.3104	6	202.7777	6	283.2451	6	363.7125
7	43.4524	7	123.9197	7	204.3871	7	284.8545	7	365.3218
8	45.0617	8	125.5291	8	205.9954	8	286.4638	8	366.9312
9	46.6711	9	127.1384	9	207.6058	9	288.0732	9	368.5405
30	48.2804	80	128.7478	130	209.2151	180	289.6825	230	370.1499
1	49.8898	1	130.3571	1	210.8245	1	291.2918	1	371.7592
2	51.4991	2	131.9665	2	212.4338	2	292.9012	2	373.3686
3	53.1085	3	133.5758	3	214.0432	3	294.5105	3	374.9779
4	54.7178	4	135.1852	4	215.6525	4	296.1199	4	376.5872
5	56.3272	5	136.7945	5	217.2619	5	297.7292	5	378.1966
6	57.9365	6	138.4039	6	218.8712	6	299.3386	6	379.8059
7	59.5458	7	140.0132	7	220.4806	7	300.9479	7	381.4153
8	61.1552	8	141.6226	8	222.0899	8	302.5573	8	383.0246
9	62.7645	9	143.2319	9	223.6993	9	304.1666	9	384.6340
40	64.3739	90	144.8412	140	225.3086	190	305.7760	240	386.2433
1	65.9832	1	146.4506	1	226.9180	1	307.3853	1	387.8527
2	67.5926	2	148.0599	2	228.5273	2	308.9947	2	389.4620
3	69.2019	3	149.6693	3	230.1366	3	310.6040	3	391.0714
4	70.8113	4	151.2786	4	231.7460	4	312.2134	4	392.6807
5	72.4206	5	152.8880	5	233.3553	5	313.8227	5	394.2901
6	74.0300	6	154.4973	6	234.9647	6	315.4321	6	395.8994
7	75.6393	7	156.1067	7	236.5740	7	317.0414	7	397.5088
8	77.2487	8	157.7160	8	238.1834	8	318.6507	8	399.1181
9	78.8580	9	159.3254	9	239.7927	9	320.2601	9	400.7275

IX. LENGTH—MILES TO KILOMETRES

[Reduction factor: 1 mile = 1.609347219 kilometres]

Miles	Kilo-metres	Miles	Kilo-metres	Miles	Kilo-metres	Miles	Kilo-metres	Miles	Kilo-metres
250	402.3368	.300	482.8042	350	563.2715	400	643.7389	450	724.2062
1	403.9461	1	484.4135	1	564.8809	1	645.3482	1	725.8156
2	405.5555	2	486.0229	2	566.4902	2	646.9576	2	727.4249
3	407.1648	3	487.6322	3	568.0996	3	648.5669	3	729.0343
4	408.7742	4	489.2416	4	569.7089	4	650.1763	4	730.6436
5	410.3835	5	490.8509	5	571.3183	5	651.7856	5	732.2530
6	411.9929	6	492.4602	6	572.9276	6	653.3950	6	733.8623
7	413.6022	7	494.0696	7	574.5370	7	655.0043	7	735.4717
8	415.2116	8	495.6789	8	576.1463	8	656.6137	8	737.0810
9	416.8209	9	497.2883	9	577.7557	9	658.2230	9	738.6904
260	418.4303	310	498.8976	360	579.3650	410	659.8824	460	740.2997
1	420.0396	1	500.5070	1	580.9743	1	661.4417	1	741.9091
2	421.6490	2	502.1163	2	582.5837	2	663.0511	2	743.5184
3	423.2583	3	503.7257	3	584.1930	3	664.6604	3	745.1278
4	424.8677	4	505.3350	4	585.8024	4	666.2697	4	746.7371
5	426.4770	5	506.9444	5	587.4117	5	667.8791	5	748.3465
6	428.0864	6	508.5537	6	589.0211	6	669.4884	6	749.9558
7	429.6957	7	510.1631	7	590.6304	7	671.0978	7	751.5652
8	431.3051	8	511.7724	8	592.2398	8	672.7071	8	753.1745
9	432.9144	9	513.3818	9	593.8491	9	674.3165	9	754.7838
270	434.5237	320	514.9911	370	595.4585	420	675.9258	470	756.3932
1	436.1331	1	516.6005	1	597.0678	1	677.5352	1	758.0025
2	437.7424	2	518.2098	2	598.6772	2	679.1445	2	759.6119
3	439.3518	3	519.8192	3	600.2865	3	680.7539	3	761.2212
4	440.9611	4	521.4285	4	601.8959	4	682.3632	4	762.8306
5	442.5705	5	523.0378	5	603.5052	5	683.9726	5	764.4399
6	444.1798	6	524.6472	6	605.1145	6	685.5819	6	766.0493
7	445.7892	7	526.2565	7	606.7239	7	687.1913	7	767.6586
8	447.3985	8	527.8659	8	608.3332	8	688.8006	8	769.2680
9	449.0079	9	529.4752	9	609.9426	9	690.4100	9	770.8773
280	450.6172	330	531.0846	380	611.5519	430	692.0193	480	772.4867
1	452.2266	1	532.6939	1	613.1613	1	693.6287	1	774.0960
2	453.8359	2	534.3033	2	614.7706	2	695.2380	2	775.7054
3	455.4453	3	535.9126	3	616.3800	3	696.8473	3	777.3147
4	457.0546	4	537.5220	4	617.9893	4	698.4567	4	778.9241
5	458.6640	5	539.1313	5	619.5987	5	700.0660	5	780.5334
6	460.2733	6	540.7407	6	621.2080	6	701.6754	6	782.1427
7	461.8827	7	542.3500	7	622.8174	7	703.2847	7	783.7521
8	463.4920	8	543.9594	8	624.4267	8	704.8941	8	785.3614
9	465.1013	9	545.5687	9	626.0361	9	706.5034	9	786.9708
290	466.7107	340	547.1781	390	627.6454	440	708.1128	490	788.5801
1	468.3200	1	548.7874	1	629.2548	1	709.7221	1	790.1895
2	469.9294	2	550.3967	2	630.8641	2	711.3315	2	791.7988
3	471.5387	3	552.0061	3	632.4735	3	712.9408	3	793.4082
4	473.1481	4	553.6154	4	634.0828	4	714.5502	4	795.0175
5	474.7574	5	555.2248	5	635.6922	5	716.1595	5	796.6269
6	476.3668	6	556.8341	6	637.3015	6	717.7689	6	798.2362
7	477.9761	7	558.4435	7	638.9108	7	719.3782	7	799.8456
8	479.5855	8	560.0528	8	640.5202	8	720.9876	8	801.4549
9	481.1948	9	561.6622	9	642.1295	9	722.5969	9	803.0643

IX. LENGTH—MILES TO KILOMETRES

[Reduction factor: 1 mile = 1.609347219 kilometres]

Miles	Kilometres	Miles	Kilometres	Miles	Kilometres	Miles	Kilometres	Miles	Kilometres
500	804.6736	550	885.1410	600	965.6083	650	1,046.0757	700	1,126.5431
1	806.2830	1	886.7503	1	967.2177	1	1,047.6850	1	1,128.1524
2	807.8923	2	888.3597	2	968.8270	2	1,049.2944	2	1,129.7617
3	809.5017	3	889.9690	3	970.4364	3	1,050.9037	3	1,131.3711
4	811.1110	4	891.5784	4	972.0457	4	1,052.5131	4	1,132.9804
5	812.7203	5	893.1877	5	973.6551	5	1,054.1224	5	1,134.5898
6	814.3297	6	894.7971	6	975.2644	6	1,055.7318	6	1,136.1991
7	815.9390	7	896.4064	7	976.8738	7	1,057.3411	7	1,137.8085
8	817.5484	8	898.0157	8	978.4831	8	1,058.9505	8	1,139.4178
9	819.1577	9	899.6251	9	980.0925	9	1,060.5598	9	1,141.0272
510	820.7671	560	901.2344	610	981.7018	660	1,062.1692	710	1,142.6365
1	822.3764	1	902.8438	1	983.3112	1	1,063.7785	1	1,144.2459
2	823.9858	2	904.4531	2	984.9205	2	1,065.3879	2	1,145.8552
3	825.5951	3	906.0625	3	986.5298	3	1,066.9972	3	1,147.4646
4	827.2045	4	907.6718	4	988.1392	4	1,068.6066	4	1,149.0739
5	828.8138	5	909.2812	5	989.7485	5	1,070.2159	5	1,150.6833
6	830.4232	6	910.8905	6	991.3579	6	1,071.8252	6	1,152.2926
7	832.0325	7	912.4999	7	992.9672	7	1,073.4346	7	1,153.9020
8	833.6419	8	914.1092	8	994.5766	8	1,075.0439	8	1,155.5113
9	835.2512	9	915.7186	9	996.1859	9	1,076.6533	9	1,157.1207
520	836.8606	570	917.3279	620	997.7953	670	1,078.2626	720	1,158.7300
1	838.4699	1	918.9373	1	999.4046	1	1,079.8720	1	1,160.3393
2	840.0792	2	920.5466	2	1,001.0140	2	1,081.4813	2	1,161.9487
3	841.6886	3	922.1560	3	1,002.6233	3	1,083.0907	3	1,163.5580
4	843.2979	4	923.7653	4	1,004.2327	4	1,084.7000	4	1,165.1674
5	844.9073	5	925.3747	5	1,005.8420	5	1,086.3094	5	1,166.7767
6	846.5166	6	926.9840	6	1,007.4514	6	1,087.9187	6	1,168.3861
7	848.1260	7	928.5933	7	1,009.0607	7	1,089.5281	7	1,169.9954
8	849.7353	8	930.2027	8	1,010.6701	8	1,091.1374	8	1,171.6048
9	851.3447	9	931.8120	9	1,012.2794	9	1,092.7468	9	1,173.2141
530	852.9540	580	933.4214	630	1,013.8887	680	1,094.3561	730	1,174.8235
1	854.5634	1	935.0307	1	1,015.4981	1	1,095.9655	1	1,176.4328
2	856.1727	2	936.6401	2	1,017.1074	2	1,097.5748	2	1,178.0422
3	857.7821	3	938.2494	3	1,018.7168	3	1,099.1842	3	1,179.6515
4	859.3914	4	939.8588	4	1,020.3261	4	1,100.7935	4	1,181.2609
5	861.0008	5	941.4681	5	1,021.9355	5	1,102.4028	5	1,182.8702
6	862.6101	6	943.0775	6	1,023.5448	6	1,104.0122	6	1,184.4796
7	864.2195	7	944.6868	7	1,025.1542	7	1,105.6215	7	1,186.0889
8	865.8288	8	946.2962	8	1,026.7635	8	1,107.2309	8	1,187.6982
9	867.4382	9	947.9055	9	1,028.3729	9	1,108.8402	9	1,189.3076
540	869.0475	590	949.5149	640	1,029.9822	690	1,110.4496	740	1,190.9169
1	870.6568	1	951.1242	1	1,031.5916	1	1,112.0589	1	1,192.5263
2	872.2662	2	952.7336	2	1,033.2009	2	1,113.6683	2	1,194.1356
3	873.8755	3	954.3429	3	1,034.8103	3	1,115.2776	3	1,195.7450
4	875.4849	4	955.9522	4	1,036.4196	4	1,116.8870	4	1,197.3543
5	877.0942	5	957.5616	5	1,038.0290	5	1,118.4963	5	1,198.9637
6	878.7036	6	959.1709	6	1,039.6383	6	1,120.1057	6	1,200.5730
7	880.3129	7	960.7803	7	1,041.2477	7	1,121.7150	7	1,202.1824
8	881.9223	8	962.3896	8	1,042.8570	8	1,123.3244	8	1,203.7917
9	883.5316	9	963.9990	9	1,044.4663	9	1,124.9337	9	1,205.4011

IX. LENGTH—MILES TO KILOMETRES

[Reduction factor: 1 mile = 1.609347219 kilometres]

Miles	Kilometres	Miles	Kilometres	Miles	Kilometres	Miles	Kilometres	Miles	Kilometres
750	1,207.0104	800	1,287.4778	850	1,367.9451	900	1,448.4125	950	1,528.8799
1	1,208.6198	1	1,289.0871	1	1,369.5545	1	1,450.0218	1	1,530.4892
2	1,210.2291	2	1,290.6965	2	1,371.1638	2	1,451.6312	2	1,532.0986
3	1,211.8385	3	1,292.3058	3	1,372.7732	3	1,453.2405	3	1,533.7079
4	1,213.4478	4	1,293.9152	4	1,374.3825	4	1,454.8499	4	1,535.3172
5	1,215.0572	5	1,295.5245	5	1,375.9919	5	1,456.4592	5	1,536.9266
6	1,216.6665	6	1,297.1339	6	1,377.6012	6	1,458.0686	6	1,538.5359
7	1,218.2758	7	1,298.7432	7	1,379.2106	7	1,459.6779	7	1,540.1453
8	1,219.8852	8	1,300.3526	8	1,380.8199	8	1,461.2873	8	1,541.7546
9	1,221.4945	9	1,301.9619	9	1,382.4293	9	1,462.8966	9	1,543.3640
760	1,223.1039	810	1,303.5712	860	1,384.0386	910	1,464.5060	960	1,544.9733
1	1,224.7132	1	1,305.1806	1	1,385.6480	1	1,466.1153	1	1,546.5827
2	1,226.3226	2	1,306.7899	2	1,387.2573	2	1,467.7247	2	1,548.1920
3	1,227.9319	3	1,308.3993	3	1,388.8666	3	1,469.3340	3	1,549.8014
4	1,229.5413	4	1,310.0086	4	1,390.4760	4	1,470.9434	4	1,551.4107
5	1,231.1506	5	1,311.6180	5	1,392.0853	5	1,472.5527	5	1,553.0201
6	1,232.7600	6	1,313.2273	6	1,393.6947	6	1,474.1621	6	1,554.6294
7	1,234.3693	7	1,314.8367	7	1,395.3040	7	1,475.7714	7	1,556.2388
8	1,235.9787	8	1,316.4460	8	1,396.9134	8	1,477.3807	8	1,557.8481
9	1,237.5880	9	1,318.0554	9	1,398.5227	9	1,478.9901	9	1,559.4575
770	1,239.1974	820	1,319.6647	870	1,400.1321	920	1,480.5994	970	1,561.0668
1	1,240.8067	1	1,321.2741	1	1,401.7414	1	1,482.2088	1	1,562.6761
2	1,242.4161	2	1,322.8834	2	1,403.3508	2	1,483.8181	2	1,564.2855
3	1,244.0254	3	1,324.4928	3	1,404.9601	3	1,485.4275	3	1,565.8948
⟨	1,245.6347	4	1,326.1021	4	1,406.5695	4	1,487.0368	4	1,567.5042
5	1,247.2441	5	1,327.7115	5	1,408.1788	5	1,488.6462	5	1,569.1135
6	1,248.8534	6	1,329.3208	6	1,409.7882	6	1,490.2555	6	1,570.7229
7	1,250.4628	7	1,330.9301	7	1,411.3975	7	1,491.8649	7	1,572.3322
8	1,252.0721	8	1,332.5395	8	1,413.0069	8	1,493.4742	8	1,573.9416
9	1,253.6815	9	1,334.1488	9	1,414.6162	9	1,495.0836	9	1,575.5509
780	1,255.2908	830	1,335.7582	880	1,416.2256	930	1,496.6929	980	1,577.1603
1	1,256.9002	1	1,337.3675	1	1,417.8349	1	1,498.3023	1	1,578.7696
2	1,258.5095	2	1,338.9769	2	1,419.4442	2	1,499.9116	2	1,580.3790
3	1,260.1189	3	1,340.5862	3	1,421.0536	3	1,501.5210	3	1,581.9883
4	1,261.7282	4	1,342.1956	4	1,422.6629	4	1,503.1303	4	1,583.5977
5	1,263.3376	5	1,343.8049	5	1,424.2723	5	1,504.7396	5	1,585.2070
6	1,264.9469	6	1,345.4143	6	1,425.8816	6	1,506.3490	6	1,586.8164
7	1,266.5563	7	1,347.0236	7	1,427.4910	7	1,507.9583	7	1,588.4257
8	1,268.1656	8	1,348.6330	8	1,429.1003	8	1,509.5677	8	1,590.0351
9	1,269.7750	9	1,350.2423	9	1,430.7097	9	1,511.1770	9	1,591.6444
790	1,271.3843	840	1,351.8517	890	1,432.3190	940	1,512.7864	990	1,593.2537
1	1,272.9936	1	1,353.4610	1	1,433.9284	1	1,514.3957	1	1,594.8631
2	1,274.6030	2	1,355.0704	2	1,435.5377	2	1,516.0051	2	1,596.4724
3	1,276.2123	3	1,356.6797	3	1,437.1471	3	1,517.6144	3	1,598.0818
4	1,277.8217	4	1,358.2891	4	1,438.7564	4	1,519.2238	4	1,599.6911
5	1,279.4310	5	1,359.8984	5	1,440.3658	5	1,520.8331	5	1,601.3005
6	1,281.0404	6	1,361.5077	6	1,441.9751	6	1,522.4425	6	1,602.9098
7	1,282.6497	7	1,363.1171	7	1,443.5845	7	1,524.0518	7	1,604.5192
8	1,284.2591	8	1,364.7264	8	1,445.1938	8	1,525.6612	8	1,606.1285
9	1,285.8684	9	1,366.3358	9	1,446.8031	9	1,527.2705	9	1,607.7379

IX. AREA—HECTARES TO ACRES

[Reduction factor: 1 hectare = 2.471043930 acres]

Hectares	Acres	Hectares	Acres	Hectares	Acres	Hectares	Acres	Hectares	Acres
0		50	123.55220	100	247.10439	150	370.65659	200	494.20879
1	2.47104	1	126.02324	1	249.57544	1	373.12763	1	496.67983
2	4.94209	2	128.49428	2	252.04648	2	375.59868	2	499.15087
3	7.41313	3	130.96533	3	254.51752	3	378.06972	3	501.62192
4	9.88418	4	133.43637	4	256.98857	4	380.54077	4	504.09296
5	12.35522	5	135.90742	5	259.45961	5	383.01181	5	506.56401
6	14.82626	6	138.37846	6	261.93066	6	385.48285	6	509.03505
7	17.29731	7	140.84950	7	264.40170	7	387.95390	7	511.50609
8	19.76835	8	143.33055	8	266.87274	8	390.42494	8	513.97714
9	22.23940	9	145.79159	9	269.34379	9	392.89598	9	516.44818
10	24.71044	60	148.26264	110	271.81483	160	395.36703	210	518.91923
1	27.18148	1	150.73368	1	274.28588	1	397.83807	1	521.39027
2	29.65253	2	153.20472	2	276.75692	2	400.30912	2	523.86131
3	32.12357	3	155.67577	3	279.22796	3	402.78016	3	526.33236
4	34.59462	4	158.14681	4	281.69901	4	405.25120	4	528.80340
5	37.06566	5	160.61786	5	284.17005	5	407.72225	5	531.27444
6	39.53670	6	163.08890	6	286.64110	6	410.19329	6	533.74549
7	42.00775	7	165.55994	7	289.11214	7	412.66434	7	536.21653
8	44.47879	8	168.03099	8	291.58318	8	415.13538	8	538.68758
9	46.94983	9	170.50203	9	294.05423	9	417.60642	9	541.15862
20	49.42088	70	172.97308	120	296.52527	170	420.07747	220	543.62966
1	51.89192	1	175.44412	1	298.99632	1	422.54851	1	546.10071
2	54.36297	2	177.91516	2	301.46736	2	425.01956	2	548.57175
3	56.83401	3	180.38621	3	303.93840	3	427.49060	3	551.04280
4	59.30505	4	182.85725	4	306.40945	4	429.96164	4	553.51384
5	61.77610	5	185.32829	5	308.88049	5	432.43269	5	555.98488
6	64.24714	6	187.79934	6	311.35154	6	434.90373	6	558.45593
7	66.71819	7	190.27038	7	313.82258	7	437.37478	7	560.92697
8	69.18923	8	192.74143	8	316.29362	8	439.84582	8	563.39802
9	71.66027	9	195.21247	9	318.76467	9	442.31686	9	565.86906
30	74.13132	80	197.68351	130	321.23571	180	444.78791	230	568.34010
1	76.60236	1	200.15456	1	323.70675	1	447.25895	1	570.81115
2	79.07341	2	202.62560	2	326.17780	2	449.73000	2	573.28219
3	81.54445	3	205.09665	3	328.64884	3	452.20104	3	575.75324
4	84.01549	4	207.56769	4	331.11989	4	454.67208	4	578.22428
5	86.48654	5	210.03873	5	333.59093	5	457.14313	5	580.69532
6	88.95758	6	212.50978	6	336.06197	6	459.61417	6	583.16637
7	91.42863	7	214.98082	7	338.53302	7	462.08521	7	585.63741
8	93.89967	8	217.45187	8	341.00406	8	464.55626	8	588.10846
9	96.37071	9	219.92291	9	343.47511	9	467.02730	9	590.57950
40	98.84176	90	222.39395	140	345.94615	190	469.49835	240	593.05054
1	101.31280	1	224.86500	1	348.41719	1	471.96939	1	595.52159
2	103.78385	2	227.33604	2	350.88824	2	474.44043	2	597.99263
3	106.25489	3	229.80709	3	353.35928	3	476.91148	3	600.46367
4	108.72593	4	232.27813	4	355.83033	4	479.38252	4	602.93472
5	111.19698	5	234.74917	5	358.30137	5	481.85357	5	605.40576
6	113.66802	6	237.22022	6	360.77241	6	484.32461	6	607.87681
7	116.13906	7	239.69126	7	363.24346	7	486.79565	7	610.34785
8	118.61011	8	242.16231	8	365.71450	8	489.26670	8	612.81889
9	121.08115	9	244.63335	9	368.18555	9	491.73774	9	615.28994

IX. AREA—HECTARES TO ACRES

[Reduction factor: 1 hectare = 2.471043930 acres]

Hectares	Acres	Hectares	Acres	Hectares	Acres	Hectares	Acres	Hectares	Acres
250	617.76098	300	741.31318	350	864.86538	400	988.41757	450	1,111.96977
1	620.23203	1	743.78422	1	867.33642	1	990.88862	1	1,114.44081
2	622.70307	2	746.25527	2	869.80746	2	993.35966	2	1,116.91186
3	625.17411	3	748.72631	3	872.27851	3	995.83070	3	1,119.38290
4	627.64516	4	751.19735	4	874.74955	4	998.30175	4	1,121.85394
5	630.11620	5	753.66840	5	877.22060	5	1,000.77279	5	1,124.32499
6	632.58725	6	756.13944	6	879.69164	6	1,003.24384	6	1,126.79603
7	635.05829	7	758.61049	7	882.16268	7	1,005.71488	7	1,129.26708
8	637.52933	8	761.08153	8	884.63373	8	1,008.18592	8	1,131.73812
9	640.00038	9	763.55257	9	887.10477	9	1,010.65697	9	1,134.20916
260	642.47142	310	766.02362	360	889.57581	410	1,013.12801	460	1,136.68021
1	644.94247	1	768.49466	1	892.04686	1	1,015.59906	1	1,139.15125
2	647.41351	2	770.96571	2	894.51790	2	1,018.07010	2	1,141.62230
3	649.88455	3	773.43675	3	896.98895	3	1,020.54114	3	1,144.09334
4	652.35560	4	775.90779	4	899.45999	4	1,023.01219	4	1,146.56438
5	654.82664	5	778.37884	5	901.93103	5	1,025.48323	5	1,149.03543
6	657.29769	6	780.84988	6	904.40208	6	1,027.95427	6	1,151.50647
7	659.76873	7	783.32093	7	906.87312	7	1,030.42532	7	1,153.97752
8	662.23977	8	785.79197	8	909.34417	8	1,032.89636	8	1,156.44856
9	664.71082	9	788.26301	9	911.81521	9	1,035.36741	9	1,158.91960
270	667.18186	320	790.73406	370	914.28625	420	1,037.83845	470	1,161.39065
1	669.65291	1	793.20510	1	916.75730	1	1,040.30949	1	1,163.86169
2	672.12395	2	795.67615	2	919.22834	2	1,042.78054	2	1,166.33273
3	674.59499	3	798.14719	3	921.69939	3	1,045.25158	3	1,168.80378
4	677.06604	4	800.61823	4	924.17043	4	1,047.72263	4	1,171.27482
5	679.53708	5	803.08928	5	926.64147	5	1,050.19367	5	1,173.74587
6	682.00812	6	805.56032	6	929.11252	6	1,052.66471	6	1,176.21691
7	684.47917	7	808.03137	7	931.58356	7	1,055.13576	7	1,178.68795
8	686.95021	8	810.50241	8	934.05461	8	1,057.60680	8	1,181.15900
9	689.42126	9	812.97345	9	936.52565	9	1,060.07785	9	1,183.63004
280	691.89230	330	815.44450	380	938.99669	430	1,062.54889	480	1,186.10109
1	694.36334	1	817.91554	1	941.46774	1	1,065.01993	1	1,188.57213
2	696.83439	2	820.38658	2	943.93878	2	1,067.49098	2	1,191.04317
3	699.30543	3	822.85763	3	946.40983	3	1,069.96202	3	1,193.51422
4	701.77648	4	825.32867	4	948.88087	4	1,072.43307	4	1,195.98526
5	704.24752	5	827.79972	5	951.35191	5	1,074.90411	5	1,198.45631
6	706.71856	6	830.27076	6	953.82296	6	1,077.37515	6	1,200.92735
7	709.18961	7	832.74180	7	956.29400	7	1,079.84620	7	1,203.39839
8	711.66065	8	835.21285	8	958.76504	8	1,082.31724	8	1,205.86944
9	714.13170	9	837.68389	9	961.23609	9	1,084.78829	9	1,208.34048
290	716.60274	340	840.15494	390	963.70713	440	1,087.25933	490	1,210.81153
1	719.07378	1	842.62598	1	966.17818	1	1,089.73037	1	1,213.28257
2	721.54483	2	845.09702	2	968.64922	2	1,092.20142	2	1,215.75361
3	724.01587	3	847.56807	3	971.12026	3	1,094.67246	3	1,218.22466
4	726.48692	4	850.03911	4	973.59131	4	1,097.14350	4	1,220.69570
5	728.95796	5	852.51016	5	976.06235	5	1,099.61455	5	1,223.16675
6	731.42900	6	854.98120	6	978.53340	6	1,102.08559	6	1,225.63779
7	733.90005	7	857.45224	7	981.00444	7	1,104.55664	7	1,228.10883
8	736.37109	8	859.92329	8	983.47548	8	1,107.02768	8	1,230.57988
9	738.84214	9	862.39433	9	985.94653	9	1,109.49872	9	1,233.05092

IX. AREA—HECTARES TO ACRES

[Reduction factor: 1 hectare = 2.471043930 acres]

Hectares	Acres	Hectares	Acres	Hectares	Acres	Hectares	Acres	Hectares	Acres
500	1,235.52197	550	1,359.07416	600	1,482.62636	650	1,606.17855	700	1,729.73075
1	1,237.99301	1	1,361.54521	1	1,485.09740	1	1,608.64960	1	1,732.20180
2	1,240.46405	2	1,364.01625	2	1,487.56845	2	1,611.12064	2	1,734.67284
3	1,242.93510	3	1,366.48729	3	1,490.03949	3	1,613.59169	3	1,737.14388
4	1,245.40614	4	1,368.95834	4	1,492.51053	4	1,616.06273	4	1,739.61493
5	1,247.87718	5	1,371.42938	5	1,494.98158	5	1,618.53377	5	1,742.08597
6	1,250.34823	6	1,373.90043	6	1,497.45262	6	1,621.00482	6	1,744.55701
7	1,252.81927	7	1,376.37147	7	1,499.92367	7	1,623.47586	7	1,747.02806
8	1,255.29032	8	1,378.84251	8	1,502.39471	8	1,625.94691	8	1,749.49910
9	1,257.76136	9	1,381.31356	9	1,504.86575	9	1,628.41795	9	1,751.97015
510	1,260.23240	560	1,383.78460	610	1,507.33680	660	1,630.88899	710	1,754.44119
1	1,262.70345	1	1,386.25564	1	1,509.80784	1	1,633.36004	1	1,756.91223
2	1,265.17449	2	1,388.72669	2	1,512.27889	2	1,635.83108	2	1,759.38328
3	1,267.64554	3	1,391.19773	3	1,514.74993	3	1,638.30213	3	1,761.85432
4	1,270.11658	4	1,393.66878	4	1,517.22097	4	1,640.77317	4	1,764.32537
5	1,272.58762	5	1,396.13982	5	1,519.69202	5	1,643.24421	5	1,766.79641
6	1,275.05867	6	1,398.61086	6	1,522.16306	6	1,645.71526	6	1,769.26745
7	1,277.52971	7	1,401.08191	7	1,524.63411	7	1,648.18630	7	1,771.73850
8	1,280.00076	8	1,403.55295	8	1,527.10515	8	1,650.65735	8	1,774.20954
9	1,282.47180	9	1,406.02400	9	1,529.57619	9	1,653.12839	9	1,776.68059
520	1,284.94284	570	1,408.49504	620	1,532.04724	670	1,655.59943	720	1,779.15163
1	1,287.41389	1	1,410.96608	1	1,534.51828	1	1,658.07048	1	1,781.62267
2	1,289.88493	2	1,413.43713	2	1,536.98932	2	1,660.54152	2	1,784.09372
3	1,292.35598	3	1,415.90817	3	1,539.46037	3	1,663.01257	3	1,786.56476
4	1,294.82702	4	1,418.37922	4	1,541.93141	4	1,665.48361	4	1,789.03581
5	1,297.29806	5	1,420.85026	5	1,544.40246	5	1,667.95465	5	1,791.50685
6	1,299.76911	6	1,423.32130	6	1,546.87350	6	1,670.42570	6	1,793.97789
7	1,302.24015	7	1,425.79235	7	1,549.34454	7	1,672.89674	7	1,796.44894
8	1,304.71120	8	1,428.26339	8	1,551.81559	8	1,675.36778	8	1,798.91998
9	1,307.18224	9	1,430.73444	9	1,554.28663	9	1,677.83883	9	1,801.39103
530	1,309.65328	580	1,433.20548	630	1,556.75768	680	1,680.30987	730	1,803.86207
1	1,312.12433	1	1,435.67652	1	1,559.22872	1	1,682.78092	1	1,806.33311
2	1,314.59537	2	1,438.14757	2	1,561.69976	2	1,685.25196	2	1,808.80416
3	1,317.06641	3	1,440.61861	3	1,564.17081	3	1,687.72300	3	1,811.27520
4	1,319.53746	4	1,443.08966	4	1,566.64185	4	1,690.19405	4	1,813.74624
5	1,322.00850	5	1,445.56070	5	1,569.11290	5	1,692.66509	5	1,816.21729
6	1,324.47955	6	1,448.03174	6	1,571.58394	6	1,695.13614	6	1,818.68833
7	1,326.95059	7	1,450.50279	7	1,574.05498	7	1,697.60718	7	1,821.15938
8	1,329.42163	8	1,452.97383	8	1,576.52603	8	1,700.07822	8	1,823.63042
9	1,331.89268	9	1,455.44487	9	1,578.99707	9	1,702.54927	9	1,826.10146
540	1,334.36372	590	1,457.91592	640	1,581.46812	690	1,705.02031	740	1,828.57251
1	1,336.83477	1	1,460.38696	1	1,583.93916	1	1,707.49136	1	1,831.04355
2	1,339.30581	2	1,462.85801	2	1,586.41020	2	1,709.96240	2	1,833.51460
3	1,341.77685	3	1,465.32905	3	1,588.88125	3	1,712.43344	3	1,835.98564
4	1,344.24790	4	1,467.80009	4	1,591.35229	4	1,714.90449	4	1,838.45668
5	1,346.71894	5	1,470.27114	5	1,593.82334	5	1,717.37553	5	1,840.92773
6	1,349.18999	6	1,472.74218	6	1,596.29438	6	1,719.84658	6	1,843.39877
7	1,351.66103	7	1,475.21323	7	1,598.76542	7	1,722.31762	7	1,845.86982
8	1,354.13207	8	1,477.68427	8	1,601.23647	8	1,724.78866	8	1,848.34086
9	1,356.60312	9	1,480.15531	9	1,603.70751	9	1,727.25971	9	1,850.81190

IX. AREA—HECTARES TO ACRES

[Reduction factor: 1 hectare = 2.471043930 acres]

Hectares	Acres	Hectares	Acres	Hectares	Acres	Hectares	Acres	Hectares	Acres
750	1,853.28295	800	1,976.83514	850	2,100.38734	900	2,223.93954	950	2,347.49173
1	1,855.75399	1	1,979.30619	1	2,102.85838	1	2,226.41058	1	2,349.96278
2	1,858.22504	2	1,981.77723	2	2,105.32943	2	2,228.88163	2	2,352.43382
3	1,860.69608	3	1,984.24828	3	2,107.80047	3	2,231.35267	3	2,354.90487
4	1,863.16712	4	1,986.71932	4	2,110.27152	4	2,233.82371	4	2,357.37591
5	1,865.63817	5	1,989.19036	5	2,112.74256	5	2,236.29476	5	2,359.84695
6	1,868.10921	6	1,991.66141	6	2,115.21360	6	2,238.76580	6	2,362.31800
7	1,870.58026	7	1,994.13245	7	2,117.68465	7	2,241.23684	7	2,364.78904
8	1,873.05130	8	1,996.60350	8	2,120.15569	8	2,243.70789	8	2,367.26009
9	1,875.52234	9	1,999.07454	9	2,122.62674	9	2,246.17893	9	2,369.73113
760	1,877.99339	810	2,001.54558	860	2,125.09778	910	2,248.64998	960	2,372.20217
1	1,880.46443	1	2,004.01663	1	2,127.56882	1	2,251.12102	1	2,374.67322
2	1,882.93547	2	2,006.48767	2	2,130.03987	2	2,253.59206	2	2,377.14426
3	1,885.40652	3	2,008.95872	3	2,132.51091	3	2,256.06311	3	2,379.61530
4	1,887.87756	4	2,011.42976	4	2,134.98196	4	2,258.53415	4	2,382.08635
5	1,890.34861	5	2,013.90080	5	2,137.45300	5	2,261.00520	5	2,384.55739
6	1,892.81965	6	2,016.37185	6	2,139.92404	6	2,263.47624	6	2,387.02844
7	1,895.29069	7	2,018.84289	7	2,142.39509	7	2,265.94728	7	2,389.49948
8	1,897.76174	8	2,021.31394	8	2,144.86613	8	2,268.41833	8	2,391.97052
9	1,900.23278	9	2,023.78498	9	2,147.33718	9	2,270.88937	9	2,394.44157
770	1,902.70383	820	2,026.25602	870	2,149.80822	920	2,273.36042	970	2,396.91261
1	1,905.17487	1	2,028.72707	1	2,152.27926	1	2,275.83146	1	2,399.38366
2	1,907.64591	2	2,031.19811	2	2,154.75031	2	2,278.30250	2	2,401.85470
3	1,910.11696	3	2,033.66915	3	2,157.22135	3	2,280.77355	3	2,404.32574
4	1,912.58800	4	2,036.14020	4	2,159.69240	4	2,283.24459	4	2,406.79679
5	1,915.05905	5	2,038.61124	5	2,162.16344	5	2,285.71564	5	2,409.26783
6	1,917.53009	6	2,041.08229	6	2,164.63448	6	2,288.18668	6	2,411.73888
7	1,920.00113	7	2,043.55333	7	2,167.10553	7	2,290.65772	7	2,414.20992
8	1,922.47218	8	2,046.02437	8	2,169.57657	8	2,293.12877	8	2,416.68096
9	1,924.94322	9	2,048.49542	9	2,172.04761	9	2,295.59981	9	2,419.15201
780	1,927.41427	830	2,050.96646	880	2,174.51866	930	2,298.07086	980	2,421.62305
1	1,929.88531	1	2,053.43751	1	2,176.98970	1	2,300.54190	1	2,424.09410
2	1,932.35635	2	2,055.90855	2	2,179.46075	2	2,303.01294	2	2,426.56514
3	1,934.82740	3	2,058.37959	3	2,181.93179	3	2,305.48399	3	2,429.03618
4	1,937.29844	4	2,060.85064	4	2,184.40283	4	2,307.95503	4	2,431.50723
5	1,939.76949	5	2,063.32168	5	2,186.87388	5	2,310.42607	5	2,433.97827
6	1,942.24053	6	2,065.79273	6	2,189.34492	6	2,312.89712	6	2,436.44932
7	1,944.71157	7	2,068.26377	7	2,191.81597	7	2,315.36816	7	2,438.92036
8	1,947.18262	8	2,070.73481	8	2,194.28701	8	2,317.83921	8	2,441.39140
9	1,949.65366	9	2,073.20586	9	2,196.75805	9	2,320.31025	9	2,443.86245
790	1,952.12471	840	2,075.67690	890	2,199.22910	940	2,322.78129	990	2,446.33349
1	1,954.59575	1	2,078.14795	1	2,201.70014	1	2,325.25234	1	2,448.80454
2	1,957.06679	2	2,080.61899	2	2,204.17119	2	2,327.72338	2	2,451.27558
3	1,959.53784	3	2,083.09003	3	2,206.64223	3	2,330.19443	3	2,453.74662
4	1,962.00888	4	2,085.56108	4	2,209.11327	4	2,332.66547	4	2,456.21767
5	1,964.47992	5	2,088.03212	5	2,211.58432	5	2,335.13651	5	2,458.68871
6	1,966.95097	6	2,090.50317	6	2,214.05536	6	2,337.60756	6	2,461.15975
7	1,969.42201	7	2,092.97421	7	2,216.52641	7	2,340.07860	7	2,463.63080
8	1,971.89306	8	2,095.44525	8	2,218.99745	8	2,342.54965	8	2,466.10184
9	1,974.36410	9	2,097.91630	9	2,221.46849	9	2,345.02069	9	2,468.57289

IX. AREA—ACRES TO HECTARES

[Reduction factor: 1 acre = 0.4046872610 hectare]

Acres	Hectares	Acres	Hectares	Acres	Hectares	Acres	Hectares	Acres	Hectares
0		50	20.23436	100	40.46873	150	60.70309	200	80.93745
1	0.40469	1	20.63905	1	40.87341	1	61.10778	1	81.34214
2	0.80937	2	21.04374	2	41.27810	2	61.51246	2	81.74683
3	1.21406	3	21.44842	3	41.68279	3	61.91715	3	82.15151
4	1.61875	4	21.85311	4	42.08748	4	62.32184	4	82.55620
5	2.02344	5	22.25780	5	42.49216	5	62.72653	5	82.96089
6	2.42812	6	22.66249	6	42.89685	6	63.13121	6	83.36558
7	2.83281	7	23.06717	7	43.30154	7	63.53590	7	83.77026
8	3.23750	8	23.47186	8	43.70622	8	63.94059	8	84.17495
9	3.64219	9	23.87655	9	44.11091	9	64.34527	9	84.57964
10	4.04687	60	24.28124	110	44.51560	160	64.74996	210	84.98432
1	4.45156	1	24.68592	1	44.92029	1	65.15465	1	85.38901
2	4.85625	2	25.09061	2	45.32497	2	65.55934	2	85.79370
3	5.26093	3	25.49530	3	45.72966	3	65.96402	3	86.19839
4	5.66562	4	25.89998	4	46.13435	4	66.36871	4	86.60307
5	6.07031	5	26.30467	5	46.53904	5	66.77340	5	87.00776
6	6.47500	6	26.70936	6	46.94372	6	67.17809	6	87.41245
7	6.87968	7	27.11405	7	47.34841	7	67.58277	7	87.81714
8	7.28437	8	27.51873	8	47.75310	8	67.98746	8	88.22182
9	7.68906	9	27.92342	9	48.15778	9	68.39215	9	88.62651
20	8.09375	70	28.32811	120	48.56247	170	68.79683	220	89.03120
1	8.49843	1	28.73280	1	48.96716	1	69.20152	1	89.43588
2	8.90312	2	29.13748	2	49.37185	2	69.60621	2	89.84057
3	9.30781	3	29.54217	3	49.77653	3	70.01090	3	90.24526
4	9.71249	4	29.94686	4	50.18122	4	70.41558	4	90.64995
5	10.11718	5	30.35154	5	50.58591	5	70.82027	5	91.05463
6	10.52187	6	30.75623	6	50.99059	6	71.22496	6	91.45932
7	10.92656	7	31.16092	7	51.39528	7	71.62965	7	91.86401
8	11.33124	8	31.56561	8	51.79997	8	72.03433	8	92.26870
9	11.73593	9	31.97029	9	52.20466	9	72.43902	9	92.67338
30	12.14062	80	32.37498	130	52.60934	180	72.84371	230	93.07807
1	12.54531	1	32.77967	1	53.01403	1	73.24839	1	93.48276
2	12.94999	2	33.18436	2	53.41872	2	73.65308	2	93.88744
3	13.35468	3	33.58904	3	53.82341	3	74.05777	3	94.29213
4	13.75937	4	33.99373	4	54.22809	4	74.46246	4	94.69682
5	14.16405	5	34.39842	5	54.63278	5	74.86714	5	95.10151
6	14.56874	6	34.80310	6	55.03747	6	75.27183	6	95.50619
7	14.97343	7	35.20779	7	55.44215	7	75.67652	7	95.91088
8	15.37812	8	35.61248	8	55.84684	8	76.08121	8	96.31557
9	15.78280	9	36.01717	9	56.25153	9	76.48589	9	96.72026
40	16.18749	90	36.42185	140	56.65622	190	76.89058	240	97.12494
1	16.59218	1	36.82654	1	57.06090	1	77.29527	1	97.52963
2	16.99686	2	37.23123	2	57.46559	2	77.69995	2	97.93432
3	17.40155	3	37.63592	3	57.87028	3	78.10464	3	98.33900
4	17.80624	4	38.04060	4	58.27497	4	78.50933	4	98.74369
5	18.21093	5	38.44529	5	58.67965	5	78.91402	5	99.14838
6	18.61561	6	38.84998	6	59.08434	6	79.31870	6	99.55307
7	19.02030	7	39.25466	7	59.48903	7	79.72339	7	99.95775
8	19.42499	8	39.65935	8	59.89371	8	80.12808	8	100.36244
9	19.82968	9	40.06404	9	60.29840	9	80.53276	9	100.76713

IX. AREA—ACRES TO HECTARES

[Reduction factor: 1 acre = 0.4046872610 hectare]

Acres	Hectares	Acres	Hectares	Acres	Hectares	Acres	Hectares	Acres	Hectares
250	101.17182	300	121.40618	350	141.64054	400	161.87490	450	182.10927
1	101.57650	1	121.81087	1	142.04523	1	162.27959	1	182.51395
2	101.98119	2	122.21555	2	142.44992	2	162.68428	2	182.91864
3	102.38588	3	122.62024	3	142.85460	3	163.08897	3	183.32333
4	102.79056	4	123.02493	4	143.25929	4	163.49365	4	183.72802
5	103.19525	5	123.42961	5	143.66398	5	163.89834	5	184.13270
6	103.59994	6	123.83430	6	144.06866	6	164.30303	6	184.53739
7	104.00463	7	124.23899	7	144.47335	7	164.70772	7	184.94208
8	104.40931	8	124.64368	8	144.87804	8	165.11240	8	185.34677
9	104.81400	9	125.04836	9	145.28273	9	165.51709	9	185.75145
260	105.21869	310	125.45305	360	145.68741	410	165.92178	460	186.15614
1	105.62338	1	125.85774	1	146.09210	1	166.32646	1	186.56083
2	106.02806	2	126.26243	2	146.49679	2	166.73115	2	186.96551
3	106.43275	3	126.66711	3	146.90148	3	167.13584	3	187.37020
4	106.83744	4	127.07180	4	147.30616	4	167.54053	4	187.77489
5	107.24212	5	127.47649	5	147.71085	5	167.94521	5	188.17958
6	107.64681	6	127.88117	6	148.11554	6	168.34990	6	188.58426
7	108.05150	7	128.28586	7	148.52022	7	168.75459	7	188.98895
8	108.45619	8	128.69055	8	148.92491	8	169.15928	8	189.39364
9	108.86087	9	129.09524	9	149.32960	9	169.56396	9	189.79833
270	109.26556	320	129.49992	370	149.73429	420	169.96865	470	190.20301
1	109.67025	1	129.90461	1	150.13897	1	170.37334	1	190.60770
2	110.07493	2	130.30930	2	150.54366	2	170.77802	2	191.01239
3	110.47962	3	130.71399	3	150.94835	3	171.18271	3	191.41707
4	110.88431	4	131.11867	4	151.35304	4	171.58740	4	191.82176
5	111.28900	5	131.52336	5	151.75772	5	171.99209	5	192.22645
6	111.69368	6	131.92805	6	152.16241	6	172.39677	6	192.63114
7	112.09837	7	132.33273	7	152.56710	7	172.80146	7	193.03582
8	112.50306	8	132.73742	8	152.97178	8	173.20615	8	193.44051
9	112.90775	9	133.14211	9	153.37647	9	173.61083	9	193.84520
280	113.31243	330	133.54680	380	153.78116	430	174.01552	480	194.24989
1	113.71712	1	133.95148	1	154.18585	1	174.42021	1	194.65457
2	114.12181	2	134.35617	2	154.59053	2	174.82490	2	195.05926
3	114.52649	3	134.76086	3	154.99522	3	175.22958	3	195.46395
4	114.93118	4	135.16555	4	155.39991	4	175.63427	4	195.86863
5	115.33587	5	135.57023	5	155.80460	5	176.03896	5	196.27332
6	115.74056	6	135.97492	6	156.20928	6	176.44365	6	196.67801
7	116.14524	7	136.37961	7	156.61397	7	176.84833	7	197.08270
8	116.54993	8	136.78429	8	157.01866	8	177.25302	8	197.48738
9	116.95462	9	137.18898	9	157.42334	9	177.65771	9	197.89207
290	117.35931	340	137.59367	390	157.82803	440	178.06239	490	198.29676
1	117.76399	1	137.99836	1	158.23272	1	178.46708	1	198.70145
2	118.16868	2	138.40304	2	158.63741	2	178.87177	2	199.10613
3	118.57337	3	138.80773	3	159.04209	3	179.27646	3	199.51082
4	118.97805	4	139.21242	4	159.44678	4	179.68114	4	199.91551
5	119.38274	5	139.61711	5	159.85147	5	180.08583	5	200.32019
6	119.78743	6	140.02179	6	160.25616	6	180.49052	6	200.72488
7	120.19212	7	140.42648	7	160.66084	7	180.89521	7	201.12957
8	120.59680	8	140.83117	8	161.06553	8	181.29989	8	201.53426
9	121.00149	9	141.23585	9	161.47022	9	181.70458	9	201.93894

IX. AREA—ACRES TO HECTARES

[Reduction factor: 1 acre = 0.4046872610 hectare]

Acres	Hectares	Acres	Hectares	Acres	Hectares	Acres	Hectares	Acres	Hectares
500	202. 34363	550	222. 57799	600	242. 81236	650	263. 04672	700	283. 28108
1	202. 74832	1	222. 98268	1	243. 21704	1	263. 45141	1	283. 68577
2	203. 15301	2	223. 38737	2	243. 62173	2	263. 85609	2	284. 09046
3	203. 55769	3	223. 79206	3	244. 02642	3	264. 26078	3	284. 49514
4	203. 96238	4	224. 19674	4	244. 43111	4	264. 66547	4	284. 89983
5	204. 36707	5	224. 60143	5	244. 83579	5	265. 07016	5	285. 30452
6	204. 77175	6	225. 00612	6	245. 24048	6	265. 47484	6	285. 70921
7	205. 17644	7	225. 41080	7	245. 64517	7	265. 87953	7	286. 11389
8	205. 58113	8	225. 81549	8	246. 04985	8	266. 28422	8	286. 51858
9	205. 98582	9	226. 22018	9	246. 45454	9	266. 68890	9	286. 92327
510	206. 39050	560	226. 62487	610	246. 85923	660	267. 09359	710	287. 32796
1	206. 79519	1	227. 02955	1	247. 26392	1	267. 49828	1	287. 73264
2	207. 19988	2	227. 43424	2	247. 66860	2	267. 90297	2	288. 13733
3	207. 60456	3	227. 83893	3	248. 07329	3	268. 30765	3	288. 54202
4	208. 00925	4	228. 24362	4	248. 47798	4	268. 71234	4	288. 94670
5	208. 41394	5	228. 64830	5	248. 88267	5	269. 11703	5	289. 35139
6	208. 81863	6	229. 05299	6	249. 28735	6	269. 52172	6	289. 75608
7	209. 22331	7	229. 45768	7	249. 69204	7	269. 92640	7	290. 16077
8	209. 62800	8	229. 86236	8	250. 09673	8	270. 33109	8	290. 56545
9	210. 03269	9	230. 26705	9	250. 50141	9	270. 73578	9	290. 97014
520	210. 43738	570	230. 67174	620	250. 90610	670	271. 14046	720	291. 37483
1	210. 84206	1	231. 07643	1	251. 31079	1	271. 54515	1	291. 77952
2	211. 24675	2	231. 48111	2	251. 71548	2	271. 94984	2	292. 18420
3	211. 65144	3	231. 88580	3	252. 12016	3	272. 35453	3	292. 58889
4	212. 05612	4	232. 29049	4	252. 52485	4	272. 75921	4	292. 99358
5	212. 46081	5	232. 69518	5	252. 92954	5	273. 16390	5	293. 39826
6	212. 86550	6	233. 09986	6	253. 33423	6	273. 56859	6	293. 80295
7	213. 27019	7	233. 50455	7	253. 73891	7	273. 97328	7	294. 20764
8	213. 67487	8	233. 90924	8	254. 14360	8	274. 37796	8	294. 61233
9	214. 07956	9	234. 31392	9	254. 54829	9	274. 78265	9	295. 01701
530	214. 48425	580	234. 71861	630	254. 95297	680	275. 18734	730	295. 42170
1	214. 88894	1	235. 12330	1	255. 35766	1	275. 59202	1	295. 82639
2	215. 29362	2	235. 52799	2	255. 76235	2	275. 99671	2	296. 23108
3	215. 69831	3	235. 93267	3	256. 16704	3	276. 40140	3	296. 63576
4	216. 10300	4	236. 33736	4	256. 57172	4	276. 80609	4	297. 04045
5	216. 50768	5	236. 74205	5	256. 97641	5	277. 21077	5	297. 44514
6	216. 91237	6	237. 14673	6	257. 38110	6	277. 61546	6	297. 84982
7	217. 31706	7	237. 55142	7	257. 78579	7	278. 02015	7	298. 25451
8	217. 72175	8	237. 95611	8	258. 19047	8	278. 42484	8	298. 65920
9	218. 12643	9	238. 36080	9	258. 59516	9	278. 82952	9	299. 06389
540	218. 53112	590	238. 76548	640	258. 99985	690	279. 23421	740	299. 46857
1	218. 93581	1	239. 17017	1	259. 40453	1	279. 63890	1	299. 87326
2	219. 34050	2	239. 57486	2	259. 80922	2	280. 04358	2	300. 27795
3	219. 74518	3	239. 97955	3	260. 21391	3	280. 44827	3	300. 68263
4	220. 14987	4	240. 38423	4	260. 61860	4	280. 85296	4	301. 08732
5	220. 55456	5	240. 78892	5	261. 02328	5	281. 25765	5	301. 49201
6	220. 95924	6	241. 19361	6	261. 42797	6	281. 66233	6	301. 89670
7	221. 36393	7	241. 59829	7	261. 83266	7	282. 06702	7	302. 30138
8	221. 76862	8	242. 00298	8	262. 23735	8	282. 47171	8	302. 70607
9	222. 17331	9	242. 40767	9	262. 64203	9	282. 87640	9	303. 11076

IX. AREA—ACRES TO HECTARES

[Reduction factor: 1 acre = 0.4046872610 hectare]

Acres	Hectares	Acres	Hectares	Acres	Hectares	Acres	Hectares	Acres	Hectares
750	303.51545	800	323.74981	850	343.98417	900	364.21853	950	384.45290
1	303.92013	1	324.15450	1	344.38886	1	364.62322	1	384.85759
2	304.32482	2	324.55918	2	344.79355	2	365.02791	2	385.26227
3	304.72951	3	324.96387	3	345.19823	3	365.43260	3	385.66696
4	305.13419	4	325.36856	4	345.60292	4	365.83728	4	386.07165
5	305.53888	5	325.77325	5	346.00761	5	366.24197	5	386.47633
6	305.94357	6	326.17793	6	346.41230	6	366.64666	6	386.88102
7	306.34826	7	326.58262	7	346.81698	7	367.05135	7	387.28571
8	306.75294	8	326.98731	8	347.22167	8	367.45603	8	387.69040
9	307.15763	9	327.39199	9	347.62636	9	367.86072	9	388.09508
760	307.56232	810	327.79668	860	348.03104	910	368.26541	960	388.49977
1	307.96701	1	328.20137	1	348.43573	1	368.67009	1	388.90446
2	308.37169	2	328.60606	2	348.84042	2	369.07478	2	389.30915
3	308.77638	3	329.01074	3	349.24511	3	369.47947	3	389.71383
4	309.18107	4	329.41543	4	349.64979	4	369.88416	4	390.11852
5	309.58575	5	329.82012	5	350.05448	5	370.28884	5	390.52321
6	309.99044	6	330.22480	6	350.45917	6	370.69353	6	390.92789
7	310.39513	7	330.62949	7	350.86386	7	371.09822	7	391.33258
8	310.79982	8	331.03418	8	351.26854	8	371.50291	8	391.73727
9	311.20450	9	331.43887	9	351.67323	9	371.90759	9	392.14196
770	311.60919	820	331.84355	870	352.07792	920	372.31228	970	392.54664
1	312.01388	1	332.24824	1	352.48260	1	372.71697	1	392.95133
2	312.41857	2	332.65293	2	352.88729	2	373.12165	2	393.35602
3	312.82325	3	333.05762	3	353.29198	3	373.52634	3	393.76070
4	313.22794	4	333.46230	4	353.69667	4	373.93103	4	394.16539
5	313.63263	5	333.86699	5	354.10135	5	374.33572	5	394.57008
6	314.03731	6	334.27168	6	354.50604	6	374.74040	6	394.97477
7	314.44200	7	334.67636	7	354.91073	7	375.14509	7	395.37945
8	314.84669	8	335.08105	8	355.31542	8	375.54978	8	395.78414
9	315.25138	9	335.48574	9	355.72010	9	375.95447	9	396.18883
780	315.65606	830	335.89043	880	356.12479	930	376.35915	980	396.59352
1	316.06075	1	336.29511	1	356.52948	1	376.76384	1	396.99820
2	316.46544	2	336.69980	2	356.93416	2	377.16853	2	397.40289
3	316.87013	3	337.10449	3	357.33885	3	377.57321	3	397.80758
4	317.27481	4	337.50918	4	357.74354	4	377.97790	4	398.21226
5	317.67950	5	337.91386	5	358.14823	5	378.38259	5	398.61695
6	318.08419	6	338.31855	6	358.55291	6	378.78728	6	399.02164
7	318.48887	7	338.72324	7	358.95760	7	379.19196	7	399.42633
8	318.89356	8	339.12792	8	359.36229	8	379.59665	8	399.83101
9	319.29825	9	339.53261	9	359.76698	9	380.00134	9	400.23570
790	319.70294	840	339.93730	890	360.17166	940	380.40603	990	400.64039
1	320.10762	1	340.34199	1	360.57635	1	380.81071	1	401.04508
2	320.51231	2	340.74667	2	360.98104	2	381.21540	2	401.44976
3	320.91700	3	341.15136	3	361.38572	3	381.62009	3	401.85445
4	321.32169	4	341.55605	4	361.79041	4	382.02477	4	402.25914
5	321.72637	5	341.96074	5	362.19510	5	382.42946	5	402.66382
6	322.13106	6	342.36542	6	362.59979	6	382.83415	6	403.06851
7	322.53575	7	342.77011	7	363.00447	7	383.23884	7	403.47320
8	322.94043	8	343.17480	8	363.40916	8	383.64352	8	403.87789
9	323.34512	9	343.57948	9	363.81385	9	384.04821	9	404.28257

113

IX. VOLUME—CUBIC METRES TO CUBIC YARDS

[Reduction factor: 1 cubic metre = 1.307942772 cubic yards]

Cubic metres	Cubic yards	Cubic metres	Cubic yards	Cubic metres	Cubic yards	Cubic metres	Cubic yards	Cubic metres	Cubic yards
0		50	65.39714	100	130.79428	150	196.19142	200	261.58855
1	1.30794	1	66.70508	1	132.10222	1	197.49936	1	262.89650
2	2.61589	2	68.01302	2	133.41016	2	198.80730	2	264.20444
3	3.92383	3	69.32097	3	134.71811	3	200.11524	3	265.51238
4	5.23177	4	70.62891	4	136.02605	4	201.42319	4	266.82033
5	6.53971	5	71.93685	5	137.33399	5	202.73113	5	268.12827
6	7.84766	6	73.24480	6	138.64193	6	204.03907	6	269.43621
7	9.15560	7	74.55274	7	139.94988	7	205.34702	7	270.74415
8	10.46354	8	75.86068	8	141.25782	8	206.65496	8	272.05210
9	11.77148	9	77.16862	9	142.56576	9	207.96290	9	273.36004
10	13.07943	60	78.47657	110	143.87370	160	209.27084	210	274.66798
1	14.38737	1	79.78451	1	145.18165	1	210.57879	1	275.97592
2	15.69531	2	81.09245	2	146.48959	2	211.88673	2	277.28387
3	17.00326	3	82.40039	3	147.79753	3	213.19467	3	278.59181
4	18.31120	4	83.70834	4	149.10548	4	214.50261	4	279.89975
5	19.61914	5	85.01628	5	150.41342	5	215.81056	5	281.20770
6	20.92708	6	86.32422	6	151.72136	6	217.11850	6	282.51564
7	22.23503	7	87.63217	7	153.02930	7	218.42644	7	283.82358
8	23.54297	8	88.94011	8	154.33725	8	219.73439	8	285.13152
9	24.85091	9	90.24805	9	155.64519	9	221.04233	9	286.43947
20	26.15886	70	91.55599	120	156.95313	170	222.35027	220	287.74741
1	27.46680	1	92.86394	1	158.26108	1	223.65821	1	289.05535
2	28.77474	2	94.17188	2	159.56902	2	224.96616	2	290.36330
3	30.08268	3	95.47982	3	160.87696	3	226.27410	3	291.67124
4	31.39063	4	96.78777	4	162.18490	4	227.58204	4	292.97918
5	32.69857	5	98.09571	5	163.49285	5	228.88999	5	294.28712
6	34.00651	6	99.40365	6	164.80079	6	230.19793	6	295.59507
7	35.31445	7	100.71159	7	166.10873	7	231.50587	7	296.90301
8	36.62240	8	102.01954	8	167.41667	8	232.81381	8	298.21095
9	37.93034	9	103.32748	9	168.72462	9	234.12176	9	299.51889
30	39.23828	80	104.63542	130	170.03256	180	235.42970	230	300.82684
1	40.54623	1	105.94336	1	171.34050	1	236.73764	1	302.13478
2	41.85417	2	107.25131	2	172.64845	2	238.04558	2	303.44272
3	43.16211	3	108.55925	3	173.95639	3	239.35353	3	304.75067
4	44.47005	4	109.86719	4	175.26433	4	240.66147	4	306.05861
5	45.77800	5	111.17514	5	176.57227	5	241.96941	5	307.36655
6	47.08594	6	112.48308	6	177.88022	6	243.27736	6	308.67449
7	48.39388	7	113.79102	7	179.18816	7	244.58530	7	309.98244
8	49.70183	8	115.09896	8	180.49610	8	245.89324	8	311.29038
9	51.00977	9	116.40691	9	181.80405	9	247.20118	9	312.59832
40	52.31771	90	117.71485	140	183.11199	190	248.50913	240	313.90627
1	53.62565	1	119.02279	1	184.41993	1	249.81707	1	315.21421
2	54.93360	2	120.33074	2	185.72787	2	251.12501	2	316.52215
3	56.24154	3	121.63868	3	187.03582	3	252.43295	3	317.83009
4	57.54948	4	122.94662	4	188.34376	4	253.74090	4	319.13804
5	58.85742	5	124.25456	5	189.65170	5	255.04884	5	320.44598
6	60.16537	6	125.56251	6	190.95964	6	256.35678	6	321.75302
7	61.47331	7	126.87045	7	192.26759	7	257.66473	7	323.06186
8	62.78125	8	128.17839	8	193.57553	8	258.97267	8	324.36981
9	64.08920	9	129.48633	9	194.88347	9	260.28061	9	325.67775

IX. VOLUME—CUBIC METRES TO CUBIC YARDS

[Reduction factor: 1 cubic metre = 1.307942772 cubic yards]

Cubic metres	Cubic yards	Cubic metres	Cubic yards	Cubic metres	Cubic yards	Cubic metres	Cubic yards	Cubic metres	Cubic yards
250	326.98569	300	392.38283	350	457.77997	400	523.17711	450	588.57425
1	328.29364	1	393.69077	1	459.08791	1	524.48505	1	589.88219
2	329.60158	2	394.99872	2	460.39586	2	525.79299	2	591.19013
3	330.90952	3	396.30666	3	461.70380	3	527.10094	3	592.49808
4	332.21746	4	397.61460	4	463.01174	4	528.40888	4	593.80602
5	333.52541	5	398.92255	5	464.31968	5	529.71682	5	595.11396
6	334.83335	6	400.23049	6	465.62763	6	531.02477	6	596.42190
7	336.14129	7	401.53843	7	466.93557	7	532.33271	7	597.72985
8	337.44924	8	402.84637	8	468.24351	8	533.64065	8	599.03779
9	338.75718	9	404.15432	9	469.55146	9	534.94859	9	600.34573
260	340.06512	310	405.46226	360	470.85940	410	536.25654	460	601.65368
1	341.37306	1	406.77020	1	472.16734	1	537.56448	1	602.96162
2	342.68101	2	408.07814	2	473.47528	2	538.87242	2	604.26956
3	343.98895	3	409.38609	3	474.78323	3	540.18036	3	605.57750
4	345.29689	4	410.69403	4	476.09117	4	541.48831	4	606.88545
5	346.60483	5	412.00197	5	477.39911	5	542.79625	5	608.19339
6	347.91278	6	413.30992	6	478.70705	6	544.10419	6	609.50133
7	349.22072	7	414.61786	7	480.01500	7	545.41214	7	610.80927
8	350.52866	8	415.92580	8	481.32294	8	546.72008	8	612.11722
9	351.83661	9	417.23374	9	482.63088	9	548.02802	9	613.42516
270	353.14455	320	418.54169	370	483.93883	420	549.33596	470	614.73310
1	354.45249	1	419.84963	1	485.24677	1	550.64391	1	616.04105
2	355.76043	2	421.15757	2	486.55471	2	551.95185	2	617.34899
3	357.06838	3	422.46552	3	487.86265	3	553.25979	3	618.65693
4	358.37632	4	423.77346	4	489.17060	4	554.56774	4	619.96487
5	359.68426	5	425.08140	5	490.47854	5	555.87568	5	621.27282
6	360.99221	6	426.38934	6	491.78648	6	557.18362	6	622.58076
7	362.30015	7	427.69729	7	493.09443	7	558.49156	7	623.88870
8	363.60809	8	429.00523	8	494.40237	8	559.79951	8	625.19665
9	364.91603	9	430.31317	9	495.71031	9	561.10745	9	626.50459
280	366.22398	330	431.62111	380	497.01825	430	562.41539	480	627.81253
1	367.53192	1	432.92906	1	498.32620	1	563.72333	1	629.12047
2	368.83986	2	434.23700	2	499.63414	2	565.03128	2	630.42842
3	370.14780	3	435.54494	3	500.94208	3	566.33922	3	631.73636
4	371.45575	4	436.85289	4	502.25002	4	567.64716	4	633.04430
5	372.76369	5	438.16083	5	503.55797	5	568.95511	5	634.35224
6	374.07163	6	439.46877	6	504.86591	6	570.26305	6	635.66019
7	375.37958	7	440.77671	7	506.17385	7	571.57099	7	636.96813
8	376.68752	8	442.08466	8	507.48180	8	572.87893	8	638.27607
9	377.99546	9	443.39260	9	508.78974	9	574.18688	9	639.58402
290	379.30340	340	444.70054	390	510.09768	440	575.49482	490	640.89196
1	380.61135	1	446.00849	1	511.40562	1	576.80276	1	642.19990
2	381.91929	2	447.31643	2	512.71357	2	578.11071	2	643.50784
3	383.22723	3	448.62437	3	514.02151	3	579.41865	3	644.81579
4	384.53517	4	449.93231	4	515.32945	4	580.72659	4	646.12373
5	385.84312	5	451.24026	5	516.63739	5	582.03453	5	647.43167
6	387.15106	6	452.54820	6	517.94534	6	583.34248	6	648.73961
7	388.45900	7	453.85614	7	519.25328	7	584.65042	7	650.04756
8	389.76695	8	455.16408	8	520.56122	8	585.95836	8	651.35550
9	391.07489	9	456.47203	9	521.86917	9	587.26630	9	652.66344

[Reduction factor: 1 cubic metre = 1.307942772 cubic yards]

Cubic metres	Cubic yards	Cubic metres	Cubic yards	Cubic metres	Cubic yards	Cubic metres	Cubic yards	Cubic metres	Cubic yards
500	653.97139	550	719.36852	600	784.76566	650	850.16280	700	915.55994
1	655.27933	1	720.67647	1	786.07361	1	851.47074	1	916.86788
2	656.58727	2	721.98441	2	787.38155	2	852.77869	2	918.17583
3	657.89521	3	723.29235	3	788.68949	3	854.08663	3	919.48377
4	659.20316	4	724.60030	4	789.99743	4	855.39457	4	920.79171
5	660.51110	5	725.90824	5	791.30538	5	856.70252	5	922.09965
6	661.81904	6	727.21618	6	792.61332	6	858.01046	6	923.40760
7	663.12699	7	728.52412	7	793.92126	7	859.31840	7	924.71554
8	664.43493	8	729.83207	8	795.22921	8	860.62634	8	926.02348
9	665.74287	9	731.14001	9	796.53715	9	861.93429	9	927.33143
510	667.05081	560	732.44795	610	797.84509	660	863.24223	710	928.63937
1	668.35876	1	733.75590	1	799.15303	1	864.55017	1	929.94731
2	669.66670	2	735.06384	2	800.46098	2	865.85812	2	931.25525
3	670.97464	3	736.37178	3	801.76892	3	867.16606	3	932.56320
4	672.28258	4	737.67972	4	803.07686	4	868.47400	4	933.87114
5	673.59053	5	738.98767	5	804.38480	5	869.78194	5	935.17908
6	674.89847	6	740.29561	6	805.69275	6	871.08989	6	936.48702
7	676.20641	7	741.60355	7	807.00069	7	872.39783	7	937.79497
8	677.51436	8	742.91149	8	808.30863	8	873.70577	8	939.10291
9	678.82230	9	744.21944	9	809.61658	9	875.01371	9	940.41085
520	680.13024	570	745.52738	620	810.92452	670	876.32166	720	941.71880
1	681.43818	1	746.83532	1	812.23246	1	877.62960	1	943.02674
2	682.74613	2	748.14327	2	813.54040	2	878.93754	2	944.33468
3	684.05407	3	749.45121	3	814.84835	3	880.24549	3	945.64262
4	685.36201	4	750.75915	4	816.15629	4	881.55343	4	946.95057
5	686.66996	5	752.06709	5	817.46423	5	882.86137	5	948.25851
6	687.97790	6	753.37504	6	818.77218	6	884.16931	6	949.56645
7	689.28584	7	754.68298	7	820.08012	7	885.47726	7	950.87440
8	690.59378	8	755.99092	8	821.38806	8	886.78520	8	952.18234
9	691.90173	9	757.29886	9	822.69600	9	888.09314	9	953.49028
530	693.20967	580	758.60681	630	824.00395	680	889.40108	730	954.79822
1	694.51761	1	759.91475	1	825.31189	1	890.70903	1	956.10617
2	695.82555	2	761.22269	2	826.61983	2	892.01697	2	957.41411
3	697.13350	3	762.53064	3	827.92777	3	893.32491	3	958.72205
4	698.44144	4	763.83858	4	829.23572	4	894.63286	4	960.02999
5	699.74938	5	765.14652	5	830.54366	5	895.94080	5	961.33794
6	701.05733	6	766.45446	6	831.85160	6	897.24874	6	962.64588
7	702.36527	7	767.76241	7	833.15955	7	898.55668	7	963.95382
8	703.67321	8	769.07035	8	834.46749	8	899.86463	8	965.26177
9	704.98115	9	770.37829	9	835.77543	9	901.17257	9	966.56971
540	706.28910	590	771.68624	640	837.08337	690	902.48051	740	967.87765
1	707.59704	1	772.99418	1	838.39132	1	903.78846	1	969.18559
2	708.90498	2	774.30212	2	839.69926	2	905.09640	2	970.49354
3	710.21293	3	775.61006	3	841.00720	3	906.40434	3	971.80148
4	711.52087	4	776.91801	4	842.31515	4	907.71228	4	973.10942
5	712.82881	5	778.22595	5	843.62309	5	909.02023	5	974.41737
6	714.13675	6	779.53389	6	844.93103	6	910.32817	6	975.72531
7	715.44470	7	780.84183	7	846.23897	7	911.63611	7	977.03325
8	716.75264	8	782.14978	8	847.54692	8	912.94405	8	978.34119
9	718.06058	9	783.45772	9	848.85486	9	914.25200	9	979.64914

IX. VOLUME—CUBIC METRES TO CUBIC YARDS

[Reduction factor: 1 cubic metre = 1.307942772 cubic yards]

Cubic metres	Cubic yards	Cubic metres	Cubic yards	Cubic metres	Cubic yards	Cubic metres	Cubic yards	Cubic metres	Cubic yards
750	980.95708	800	1,046.35422	850	1,111.75136	900	1,177.14849	950	1,242.54563
1	982.26502	1	1,047.66216	1	1,113.05930	1	1,178.45644	1	1,243.85358
2	983.57296	2	1,048.97010	2	1,114.36724	2	1,179.76438	2	1,245.16152
3	984.88091	3	1,050.27805	3	1,115.67518	3	1,181.07232	3	1,246.46946
4	986.18885	4	1,051.58599	4	1,116.98313	4	1,182.38027	4	1,247.77740
5	987.49679	5	1,052.89393	5	1,118.29107	5	1,183.68821	5	1,249.08535
6	988.80474	6	1,054.20187	6	1,119.59901	6	1,184.99615	6	1,250.39329
7	990.11268	7	1,055.50982	7	1,120.90696	7	1,186.30409	7	1,251.70123
8	991.42062	8	1,056.81776	8	1,122.21490	8	1,187.61204	8	1,253.00918
9	992.72856	9	1,058.12570	9	1,123.52284	9	1,188.91998	9	1,254.31712
760	994.03651	810	1,059.43365	860	1,124.83078	910	1,190.22792	960	1,255.62506
1	995.34445	1	1,060.74159	1	1,126.13873	1	1,191.53587	1	1,256.93300
2	996.65239	2	1,062.04953	2	1,127.44667	2	1,192.84381	2	1,258.24095
3	997.96034	3	1,063.35747	3	1,128.75461	3	1,194.15175	3	1,259.54889
4	999.26828	4	1,064.66542	4	1,130.06256	4	1,195.45969	4	1,260.85683
5	1,000.57622	5	1,065.97336	5	1,131.37050	5	1,196.76764	5	1,262.16477
6	1,001.88416	6	1,067.28130	6	1,132.67844	6	1,198.07558	6	1,263.47272
7	1,003.19211	7	1,068.58924	7	1,133.98638	7	1,199.38352	7	1,264.78066
8	1,004.50005	8	1,069.89719	8	1,135.29433	8	1,200.69146	8	1,266.08860
9	1,005.80799	9	1,071.20513	9	1,136.60227	9	1,201.99941	9	1,267.39655
770	1,007.11593	820	1,072.51307	870	1,137.91021	920	1,203.30735	970	1,268.70449
1	1,008.42388	1	1,073.82102	1	1,139.21815	1	1,204.61529	1	1,270.01243
2	1,009.73182	2	1,075.12896	2	1,140.52610	2	1,205.92324	2	1,271.32037
3	1,011.03976	3	1,076.43690	3	1,141.83404	3	1,207.23118	3	1,272.62832
4	1,012.34771	4	1,077.74484	4	1,143.14198	4	1,208.53912	4	1,273.93626
5	1,013.65565	5	1,079.05279	5	1,144.44993	5	1,209.84706	5	1,275.24420
6	1,014.96359	6	1,080.36073	6	1,145.75787	6	1,211.15501	6	1,276.55215
7	1,016.27153	7	1,081.66867	7	1,147.06581	7	1,212.46295	7	1,277.86009
8	1,017.57948	8	1,082.97662	8	1,148.37375	8	1,213.77089	8	1,279.16803
9	1,018.88742	9	1,084.28456	9	1,149.68170	9	1,215.07884	9	1,280.47597
780	1,020.19536	830	1,085.59250	880	1,150.98964	930	1,216.38678	980	1,281.78392
1	1,021.50330	1	1,086.90044	1	1,152.29758	1	1,217.69472	1	1,283.09186
2	1,022.81125	2	1,088.20839	2	1,153.60552	2	1,219.00266	2	1,284.39980
3	1,024.11919	3	1,089.51633	3	1,154.91347	3	1,220.31061	3	1,285.70774
4	1,025.42713	4	1,090.82427	4	1,156.22141	4	1,221.61855	4	1,287.01569
5	1,026.73508	5	1,092.13221	5	1,157.52935	5	1,222.92649	5	1,288.32363
6	1,028.04302	6	1,093.44016	6	1,158.83730	6	1,224.23443	6	1,289.63157
7	1,029.35096	7	1,094.74810	7	1,160.14524	7	1,225.54238	7	1,290.93952
8	1,030.65890	8	1,096.05604	8	1,161.45318	8	1,226.85032	8	1,292.24746
9	1,031.96685	9	1,097.36399	9	1,162.76112	9	1,228.15826	9	1,293.55540
790	1,033.27479	840	1,098.67193	890	1,164.06907	940	1,229.46621	990	1,294.86334
1	1,034.58273	1	1,099.97987	1	1,165.37701	1	1,230.77415	1	1,296.17129
2	1,035.89068	2	1,101.28781	2	1,166.68495	2	1,232.08209	2	1,297.47923
3	1,037.19862	3	1,102.59576	3	1,167.99290	3	1,233.39003	3	1,298.78717
4	1,038.50656	4	1,103.90370	4	1,169.30084	4	1,234.69798	4	1,300.09512
5	1,039.81450	5	1,105.21164	5	1,170.60878	5	1,236.00592	5	1,301.40306
6	1,041.12245	6	1,106.51959	6	1,171.91672	6	1,237.31386	6	1,302.71100
7	1,042.43039	7	1,107.82753	7	1,173.22467	7	1,238.62181	7	1,304.01894
8	1,043.73833	8	1,109.13547	8	1,174.53261	8	1,239.92975	8	1,305.32689
9	1,045.04627	9	1,110.44341	9	1,175.84055	9	1,241.23769	9	1,306.63483

117

IX. VOLUME—CUBIC YARDS TO CUBIC METRES

[Reduction factor: 1 cubic yard = 0.7645594453 cubic metre]

Cubic yards	Cubic metres	Cubic yards	Cubic metres	Cubic yards	Cubic metres	Cubic yards	Cubic metres	Cubic yards	Cubic metres
0		50	38.22797	100	76.45594	150	114.68392	200	152.91189
1	0.76456	1	38.99253	1	77.22050	1	115.44848	1	153.67645
2	1.52912	2	39.75709	2	77.98506	2	116.21304	2	154.44101
3	2.29368	3	40.52165	3	78.74962	3	116.97760	3	155.20557
4	3.05824	4	41.28621	4	79.51418	4	117.74215	4	155.97013
5	3.82280	5	42.05077	5	80.27874	5	118.50671	5	156.73469
6	4.58736	6	42.81533	6	81.04330	6	119.27127	6	157.49925
7	5.35192	7	43.57989	7	81.80786	7	120.03583	7	158.26381
8	6.11648	8	44.34445	8	82.57242	8	120.80039	8	159.02836
9	6.88104	9	45.10901	9	83.33698	9	121.56495	9	159.79292
10	7.64559	60	45.87357	110	84.10154	160	122.32951	210	160.55748
1	8.41015	1	46.63813	1	84.86610	1	123.09407	1	161.32204
2	9.17471	2	47.40269	2	85.63066	2	123.85863	2	162.08660
3	9.93927	3	48.16725	3	86.39522	3	124.62319	3	162.85116
4	10.70383	4	48.93180	4	87.15978	4	125.38775	4	163.61572
5	11.46839	5	49.69636	5	87.92434	5	126.15231	5	164.38028
6	12.23295	6	50.46092	6	88.68890	6	126.91687	6	165.14484
7	12.99751	7	51.22548	7	89.45346	7	127.68143	7	165.90940
8	13.76207	8	51.99004	8	90.21801	8	128.44599	8	166.67396
9	14.52663	9	52.75460	9	90.98257	9	129.21055	9	167.43852
20	15.29119	70	53.51916	120	91.74713	170	129.97511	220	168.20308
1	16.05575	1	54.28372	1	92.51169	1	130.73967	1	168.96764
2	16.82031	2	55.04828	2	93.27625	2	131.50422	2	169.73220
3	17.58487	3	55.81284	3	94.04081	3	132.26878	3	170.49676
4	18.34943	4	56.57740	4	94.80537	4	133.03334	4	171.26132
5	19.11399	5	57.34196	5	95.56993	5	133.79790	5	172.02588
6	19.87855	6	58.10652	6	96.33449	6	134.56246	6	172.79043
7	20.64311	7	58.87108	7	97.09905	7	135.32702	7	173.55499
8	21.40766	8	59.63564	8	97.86361	8	136.09158	8	174.31955
9	22.17222	9	60.40020	9	98.62817	9	136.85614	9	175.08411
30	22.93678	80	61.16476	130	99.39273	180	137.62070	230	175.84867
1	23.70134	1	61.92932	1	100.15729	1	138.38526	1	176.61323
2	24.46590	2	62.69387	2	100.92185	2	139.14982	2	177.37779
3	25.23046	3	63.45843	3	101.68641	3	139.91438	3	178.14235
4	25.99502	4	64.22299	4	102.45097	4	140.67894	4	178.90691
5	26.75958	5	64.98755	5	103.21553	5	141.44350	5	179.67147
6	27.52414	6	65.75211	6	103.98008	6	142.20806	6	180.43603
7	28.28870	7	66.51667	7	104.74464	7	142.97262	7	181.20059
8	29.05326	8	67.28123	8	105.50920	8	143.73718	8	181.96515
9	29.81782	9	68.04579	9	106.27376	9	144.50174	9	182.72971
40	30.58238	90	68.81035	140	107.03832	190	145.26629	240	183.49427
1	31.34694	1	69.57491	1	107.80288	1	146.03085	1	184.25883
2	32.11150	2	70.33947	2	108.56744	2	146.79541	2	185.02339
3	32.87606	3	71.10403	3	109.33200	3	147.55997	3	185.78795
4	33.64062	4	71.86859	4	110.09656	4	148.32453	4	186.55250
5	34.40518	5	72.63315	5	110.86112	5	149.08909	5	187.31706
6	35.16973	6	73.39771	6	111.62568	6	149.85365	6	188.08162
7	35.93429	7	74.16227	7	112.39024	7	150.61821	7	188.84618
8	36.69885	8	74.92683	8	113.15480	8	151.38277	8	189.61074
9	37.46341	9	75.69139	9	113.91936	9	152.14733	9	190.37530

IX. VOLUME—CUBIC YARDS TO CUBIC METRES

[Reduction factor: 1 cubic yard = 0.7645594453 cubic metre]

Cubic yards	Cubic metres	Cubic yards	Cubic metres	Cubic yards	Cubic metres	Cubic yards	Cubic metres	Cubic yards	Cubic metres
250	191.13986	300	229.36783	350	267.59581	400	305.82378	450	344.05175
1	191.90442	1	230.13239	1	268.36037	1	306.58834	1	344.81631
2	192.66898	2	230.89695	2	269.12492	2	307.35290	2	345.58087
3	193.43354	3	231.66151	3	269.88948	3	308.11746	3	346.34543
4	194.19810	4	232.42607	4	270.65404	4	308.88202	4	347.10999
5	194.96266	5	233.19063	5	271.41860	5	309.64658	5	347.87455
6	195.72722	6	233.95519	6	272.18316	6	310.41113	6	348.63911
7	196.49178	7	234.71975	7	272.94772	7	311.17569	7	349.40367
8	197.25634	8	235.48431	8	273.71228	8	311.94025	8	350.16823
9	198.02090	9	236.24887	9	274.47684	9	312.70481	9	350.93279
260	198.78546	310	237.01343	360	275.24140	410	313.46937	460	351.69734
1	199.55002	1	237.77799	1	276.00596	1	314.23393	1	352.46190
2	200.31457	2	238.54255	2	276.77052	2	314.99849	2	353.22646
3	201.07913	3	239.30711	3	277.53508	3	315.76305	3	353.99102
4	201.84369	4	240.07167	4	278.29964	4	316.52761	4	354.75558
5	202.60825	5	240.83623	5	279.06420	5	317.29217	5	355.52014
6	203.37281	6	241.60078	6	279.82876	6	318.05673	6	356.28470
7	204.13737	7	242.36534	7	280.59332	7	318.82129	7	357.04926
8	204.90193	8	243.12990	8	281.35788	8	319.58585	8	357.81382
9	205.66649	9	243.89446	9	282.12244	9	320.35041	9	358.57838
270	206.43105	320	244.65902	370	282.88699	420	321.11497	470	359.34294
1	207.19561	1	245.42358	1	283.65155	1	321.87953	1	360.10750
2	207.96017	2	246.18814	2	284.41611	2	322.64409	2	360.87206
3	208.72473	3	246.95270	3	285.18067	3	323.40865	3	361.63662
4	209.48929	4	247.71726	4	285.94523	4	324.17320	4	362.40118
5	210.25385	5	248.48182	5	286.70979	5	324.93776	5	363.16574
6	211.01841	6	249.24638	6	287.47435	6	325.70232	6	363.93030
7	211.78297	7	250.01094	7	288.23891	7	326.46688	7	364.69486
8	212.54753	8	250.77550	8	289.00347	8	327.23144	8	365.45941
9	213.31209	9	251.54006	9	289.76803	9	327.99600	9	366.22397
280	214.07664	330	252.30462	380	290.53259	430	328.76056	480	366.98853
1	214.84120	1	253.06918	1	291.29715	1	329.52512	1	367.75309
2	215.60576	2	253.83374	2	292.06171	2	330.28968	2	368.51765
3	216.37032	3	254.59830	3	292.82627	3	331.05424	3	369.28221
4	217.13488	4	255.36285	4	293.59083	4	331.81880	4	370.04677
5	217.89944	5	256.12741	5	594.35539	5	332.58336	5	370.81133
6	218.66400	6	256.89197	6	295.11995	6	333.34792	6	371.57589
7	219.42856	7	257.65653	7	295.88451	7	334.11248	7	372.34045
8	220.19312	8	258.42109	8	296.64906	8	334.87704	8	373.10501
9	220.95768	9	259.18565	9	297.41362	9	335.64160	9	373.86957
290	221.72224	340	259.95021	390	298.17818	440	336.40616	490	374.63413
1	222.48680	1	260.71477	1	298.94274	1	337.17072	1	375.39869
2	223.25136	2	261.47933	2	299.70730	2	337.93527	2	376.16325
3	224.01592	3	262.24389	3	300.47186	3	338.69983	3	376.92781
4	224.78048	4	263.00845	4	301.23642	4	339.46439	4	377.69237
5	225.54504	5	263.77301	5	302.00098	5	340.22895	5	378.45693
6	226.30960	6	264.53757	6	302.76554	6	340.99351	6	379.22148
7	227.07416	7	265.30213	7	303.53010	7	341.75807	7	379.98604
8	227.83871	8	266.06669	8	304.29466	8	342.52263	8	380.75060
9	228.60327	9	266.83125	9	305.05922	9	343.28719	9	381.51516

IX. VOLUME—CUBIC YARDS TO CUBIC METRES

[Reduction factor: 1 cubic yard = 0.7645594453 cubic metre]

Cubic yards	Cubic metres	Cubic yards	Cubic metres	Cubic yards	Cubic metres	Cubic yards	Cubic metres	Cubic yards	Cubic metres
500	382.27972	550	420.50769	600	458.73567	650	496.96364	700	535.19161
1	383.04428	1	421.27225	1	459.50023	1	497.72820	1	535.95617
2	383.80884	2	422.03681	2	460.26479	2	498.49276	2	536.72073
3	384.57340	3	422.80137	3	461.02935	3	499.25732	3	537.48529
4	385.33796	4	423.56593	4	461.79390	4	500.02188	4	538.24985
5	386.10252	5	424.33049	5	462.55846	5	500.78644	5	539.01441
6	386.86708	6	425.09505	6	463.32302	6	501.55100	6	539.77897
7	387.63164	7	425.85961	7	464.08758	7	502.31556	7	540.54353
8	388.39620	8	426.62417	8	464.85214	8	503.08012	8	541.30809
9	389.16076	9	427.38873	9	465.61670	9	503.84467	9	542.07265
510	389.92532	560	428.15329	610	466.38126	660	504.60923	710	542.83721
1	390.68988	1	428.91785	1	467.14582	1	505.37379	1	543.60177
2	391.45444	2	429.68241	2	467.91038	2	506.13835	2	544.36633
3	392.21900	3	430.44697	3	468.67494	3	506.90291	3	545.13088
4	392.98355	4	431.21153	4	469.43950	4	507.66747	4	545.89544
5	393.74811	5	431.97609	5	470.20406	5	508.43203	5	546.66000
6	394.51267	6	432.74065	6	470.96862	6	509.19659	6	547.42456
7	395.27723	7	433.50521	7	471.73318	7	509.96115	7	548.18912
8	396.04179	8	434.26976	8	472.49774	8	510.72571	8	548.95368
9	396.80635	9	435.03432	9	473.26230	9	511.49027	9	549.71824
520	397.57091	570	435.79888	620	474.02686	670	512.25483	720	550.48280
1	398.33547	1	436.56344	1	474.79142	1	513.01939	1	551.24736
2	399.10003	2	437.32800	2	475.55597	2	513.78395	2	552.01192
3	399.86459	3	438.09256	3	476.32053	3	514.54851	3	552.77648
4	400.62915	4	438.85712	4	477.08509	4	515.31307	4	553.54104
5	401.39371	5	439.62168	5	477.84965	5	516.07763	5	554.30560
6	402.15827	6	440.38624	6	478.61421	6	516.84219	6	555.07016
7	402.92283	7	441.15080	7	479.37877	7	517.60674	7	555.83472
8	403.68739	8	441.91536	8	480.14333	8	518.37130	8	556.59928
9	404.45195	9	442.67992	9	480.90789	9	519.13586	9	557.36384
530	405.21651	580	443.44448	630	481.67245	680	519.90042	730	558.12840
1	405.98107	1	444.20904	1	482.43701	1	520.66498	1	558.89295
2	406.74562	2	444.97360	2	483.20157	2	521.42954	2	559.65751
3	407.51018	3	445.73816	3	483.96613	3	522.19410	3	560.42207
4	408.27474	4	446.50272	4	484.73069	4	522.95866	4	561.18663
5	409.03930	5	447.26728	5	485.49525	5	523.72322	5	561.95119
6	409.80386	6	448.03183	6	486.25981	6	524.48778	6	562.71575
7	410.56842	7	448.79639	7	487.02437	7	525.25234	7	563.48031
8	411.33298	8	449.56095	8	487.78893	8	526.01690	8	564.24487
9	412.09754	9	450.32551	9	488.55349	9	526.78146	9	565.00943
540	412.86210	590	451.09007	640	489.31804	690	527.54602	740	565.77399
1	413.62666	1	451.85463	1	490.08260	1	528.31058	1	566.53855
2	414.39122	2	452.61919	2	490.84716	2	529.07514	2	567.30311
3	415.15578	3	453.38375	3	491.61172	3	529.83970	3	568.06767
4	415.92034	4	454.14831	4	492.37628	4	530.60426	4	568.83223
5	416.68490	5	454.91287	5	493.14084	5	531.36881	5	569.59679
6	417.44946	6	455.67743	6	493.90540	6	532.13337	6	570.36135
7	418.21402	7	456.44199	7	494.66996	7	532.89793	7	571.12591
8	418.97858	8	457.20655	8	495.43452	8	533.66249	8	571.89047
9	419.74314	9	457.97111	9	496.19908	9	534.42705	9	572.65502

IX. VOLUME—CUBIC YARDS TO CUBIC METRES

[Reduction factor: 1 cubic yard = 0.7645594453 cubic metre]

Cubic yards	Cubic metres	Cubic yards	Cubic metres	Cubic yards	Cubic metres	Cubic yards	Cubic metres	Cubic yards	Cubic metres
750	573.41958	800	611.64756	850	649.87553	900	688.10350	950	726.33147
1	574.18414	1	612.41212	1	650.64009	1	688.86806	1	727.09603
2	574.94870	2	613.17668	2	651.40465	2	689.63262	2	727.86059
3	575.71326	3	613.94123	3	652.16921	3	690.39718	3	728.62515
4	576.47782	4	614.70579	4	652.93377	4	691.16174	4	729.38971
5	577.24238	5	615.47035	5	653.69833	5	691.92630	5	730.15427
6	578.00694	6	616.23491	6	654.46289	6	692.69086	6	730.91883
7	578.77150	7	616.99947	7	655.22744	7	693.45542	7	731.68339
8	579.53606	8	617.76403	8	655.99200	8	694.21998	8	732.44795
9	580.30062	9	618.52859	9	656.75656	9	694.98454	9	733.21251
760	581.06518	810	619.29315	860	657.52112	910	695.74910	960	733.97707
1	581.82974	1	620.05771	1	658.28568	1	696.51365	1	734.74163
2	582.59430	2	620.82227	2	659.05024	2	697.27821	2	735.50619
3	583.35886	3	621.58683	3	659.81480	3	698.04277	3	736.27075
4	584.12342	4	622.35139	4	660.57936	4	698.80733	4	737.03531
5	584.88798	5	623.11595	5	661.34392	5	699.57189	5	737.79986
6	585.65254	6	623.88051	6	662.10848	6	700.33645	6	738.56442
7	586.41709	7	624.64507	7	662.87304	7	701.10101	7	739.32898
8	587.18165	8	625.40963	8	663.63760	8	701.86557	8	740.09354
9	587.94621	9	626.17419	9	664.40216	9	702.63013	9	740.85810
770	588.71077	820	626.93875	870	665.16672	920	703.39469	970	741.62266
1	589.47533	1	627.70330	1	665.93128	1	704.15925	1	742.38722
2	590.23989	2	628.46786	2	666.69584	2	704.92381	2	743.15178
3	591.00445	3	629.23242	3	667.46040	3	705.68837	3	743.91634
4	591.76901	4	629.99698	4	668.22496	4	706.45293	4	744.68090
5	592.53357	5	630.76154	5	668.98951	5	707.21749	5	745.44546
6	593.29813	6	631.52610	6	669.75407	6	707.98205	6	746.21002
7	594.06269	7	632.29066	7	670.51863	7	708.74661	7	746.97458
8	594.82725	8	633.05522	8	671.28319	8	709.51177	8	747.73914
9	595.59181	9	633.81978	9	672.04775	9	710.27572	9	748.50370
780	596.35637	830	634.58434	880	672.81231	930	711.04028	980	749.26826
1	597.12093	1	635.34890	1	673.57687	1	711.80484	1	750.03282
2	597.88549	2	636.11346	2	674.34143	2	712.56940	2	750.79738
3	598.65005	3	636.87802	3	675.10599	3	713.33396	3	751.56193
4	599.41461	4	637.64258	4	675.87055	4	714.09852	4	752.32649
5	600.17916	5	638.40714	5	676.63511	5	714.86308	5	753.09105
6	600.94372	6	639.17170	6	677.39967	6	715.62764	6	753.85561
7	601.70828	7	639.93626	7	678.16423	7	716.39220	7	754.62017
8	602.47284	8	640.70082	8	678.92879	8	717.15676	8	755.38473
9	603.23740	9	641.46537	9	679.69335	9	717.92132	9	756.14929
790	604.00196	840	642.22993	890	680.45791	940	718.68588	990	756.91385
1	604.76652	1	642.99449	1	681.22247	1	719.45044	1	757.67841
2	605.53108	2	643.75905	2	681.98703	2	720.21500	2	758.44297
3	606.29564	3	644.52361	3	682.75158	3	720.97956	3	759.20753
4	607.06020	4	645.28817	4	683.51614	4	721.74412	4	759.97209
5	607.82476	5	646.05273	5	684.28070	5	722.50868	5	760.73665
6	608.58932	6	646.81729	6	685.04526	6	723.27324	6	761.50121
7	609.35388	7	647.58185	7	685.80982	7	724.03779	7	762.26577
8	610.11844	8	648.34641	8	686.57438	8	724.80235	8	763.03033
9	610.88300	9	649.11097	9	687.33894	9	725.56691	9	763.79489

121

IX. CAPACITY—LITRES TO LIQUID QUARTS

[Reduction factor: 1 litre = 1.0567104 quarts]

Litres	Liquid quarts	Litres	Liquid quarts	Litres	Liquid quarts	Litres	Liquid quarts	Litres	Liquid quarts
0		50	52.8355	100	105.671	150	158.507	200	211.342
1	1.0567	1	53.8922	1	106.728	1	159.563	1	212.399
2	2.1134	2	54.9489	2	107.784	2	160.620	2	213.456
3	3.1701	3	56.005?	3	108.841	3	161.677	3	214.512
4	4.2268	4	57.062+	4	109.898	4	162.733	4	215.569
5	5.2836	5	58.1191	5	110.955	5	163.790	5	216.626
6	6.3403	6	59.1758	6	112.011	6	164.847	6	217.682
7	7.3970	7	60.2325	7	113.068	7	165.904	7	218.739
8	8.4537	8	61.2892	8	114.125	8	166.960	8	219.796
9	9.5104	9	62.3459	9	115.181	9	168.017	9	220.852
10	10.5671	60	63.4026	110	116.238	160	169.074	210	221.909
1	11.6238	1	64.4593	1	117.295	1	170.130	1	222.966
2	12.6805	2	65.5160	2	118.352	2	171.187	2	224.023
3	13.7372	3	66.5728	3	119.408	3	172.244	3	225.079
4	14.7939	4	67.6295	4	120.465	4	173.301	4	226.136
5	15.8507	5	68.6862	5	121.522	5	174.357	5	227.193
6	16.9074	6	69.7429	6	122.578	6	175.414	6	228.249
7	17.9641	7	70.7996	7	123.635	7	176.471	7	229.306
8	19.0208	8	71.8563	8	124.692	8	177.527	8	230.363
9	20.0775	9	72.9130	9	125.749	9	178.584	9	231.420
20	21.1342	70	73.9697	120	126.805	170	179.641	220	232.476
1	22.1909	1	75.0264	1	127.862	1	180.697	1	233.533
2	23.2476	2	76.0831	2	128.919	2	181.754	2	234.590
3	24.3043	3	77.1399	3	129.975	3	182.811	3	235.646
4	25.3610	4	78.1966	4	131.032	4	183.868	4	236.703
5	26.4178	5	79.2533	5	132.089	5	184.924	5	237.760
6	27.4745	6	80.3100	6	133.146	6	185.981	6	238.817
7	28.5312	7	81.3667	7	134.202	7	187.038	7	239.873
8	29.5879	8	82.4234	8	135.259	8	188.094	8	240.930
9	30.6446	9	83.4801	9	136.316	9	189.151	9	241.987
30	31.7013	80	84.5368	130	137.372	180	190.208	230	243.043
1	32.7580	1	85.5935	1	138.429	1	191.265	1	244.100
2	33.8147	2	86.6503	2	139.486	2	192.321	2	245.157
3	34.8714	3	87.7070	3	140.542	3	193.378	3	246.214
4	35.9282	4	88.7637	4	141.599	4	194.435	4	247.270
5	36.9849	5	89.8204	5	142.656	5	195.491	5	248.327
6	38.0416	6	90.8771	6	143.713	6	196.548	6	249.384
7	39.0983	7	91.9338	7	144.769	7	197.605	7	250.440
8	40.1550	8	92.9905	8	145.826	8	198.662	8	251.497
9	41.2117	9	94.0472	9	146.883	9	199.718	9	252.554
40	42.2684	90	95.1039	140	147.939	190	200.775	240	253.610
1	43.3251	1	96.1606	1	148.996	1	201.832	1	254.667
2	44.3818	2	97.2174	2	150.053	2	202.888	2	255.724
3	45.4385	3	98.2741	3	151.110	3	203.945	3	256.781
4	46.4953	4	99.3308	4	152.166	4	205.002	4	257.837
5	47.5520	5	100.3875	5	153.223	5	206.059	5	258.894
6	48.6087	6	101.4442	6	154.280	6	207.115	6	259.951
7	49.6654	7	102.5009	7	155.336	7	208.172	7	261.007
8	50.7221	8	103.5576	8	156.393	8	209.229	8	262.064
9	51.7788	9	104.6143	9	157.450	9	210.285	9	263.121

IX. CAPACITY—LITRES TO LIQUID QUARTS

[Reduction factor: 1 litre = 1.0567104 quarts]

Litres	Liquid quarts	Litres	Liquid quarts	Litres	Liquid quarts	Litres	Liquid quarts	Litres	Liquid quarts
250	264.178	300	317.013	350	369.849	400	422.684	450	475.520
1	265.234	1	318.070	1	370.905	1	423.741	1	476.576
2	266.291	2	319.127	2	371.962	2	424.798	2	477.633
3	267.348	3	320.183	3	373.019	3	425.854	3	478.690
4	268.404	4	321.240	4	374.075	4	426.911	4	479.747
5	269.461	5	322.297	5	375.132	5	427.968	5	480.803
6	270.518	6	323.353	6	376.189	6	429.024	6	481.860
7	271.575	7	324.410	7	377.246	7	430.081	7	482.917
8	272.631	8	325.467	8	378.302	8	431.138	8	483.973
9	273.688	9	326.524	9	379.359	9	432.195	9	485.030
260	274.745	310	327.580	360	380.416	410	433.251	460	486.087
1	275.801	1	328.637	1	381.472	1	434.308	1	487.143
2	276.858	2	329.694	2	382.529	2	435.365	2	488.200
3	277.915	3	330.750	3	383.586	3	736.421	3	489.257
4	278.972	4	331.807	4	384.643	4	437.478	4	490.314
5	280.028	5	332.864	5	385.699	5	438.535	5	491.370
6	281.085	6	333.920	6	386.756	6	439.592	6	492.427
7	282.142	7	334.977	7	387.813	7	440.648	7	493.484
8	283.198	8	336.034	8	388.869	8	441.705	8	494.540
9	284.255	9	337.091	9	389.926	9	442.762	9	495.597
270	285.312	320	338.147	370	390.983	420	443.818	470	496.654
1	286.369	1	339.204	1	392.040	1	444.875	1	497.711
2	287.425	2	340.261	2	393.096	2	445.932	2	498.767
3	288.482	3	341.317	3	394.153	3	446.988	3	499.824
4	289.539	4	342.374	4	395.210	4	448.045	4	500.881
5	290.595	5	343.431	5	396.266	5	449.102	5	501.937
6	291.652	6	344.488	6	397.323	6	450.159	6	502.994
7	292.709	7	345.544	7	398.380	7	451.215	7	504.051
8	293.765	8	346.601	8	399.437	8	452.272	8	505.108
9	294.822	9	347.658	9	400.493	9	453.329	9	506.164
280	295.879	330	348.714	380	401.550	430	454.385	480	507.221
1	296.936	1	349.771	1	402.607	1	455.442	1	508.278
2	297.992	2	350.828	2	403.663	2	456.499	2	509.334
3	299.049	3	351.885	3	404.720	3	457.556	3	510.391
4	300.106	4	352.941	4	405.777	4	458.612	4	511.448
5	301.162	5	353.998	5	406.834	5	459.669	5	512.505
6	302.219	6	355.055	6	407.890	6	460.726	6	513.561
7	303.276	7	356.111	7	408.947	7	461.782	7	514.618
8	304.333	8	357.168	8	410.004	8	462.839	8	515.675
9	305.389	9	358.225	9	411.060	9	463.896	9	516.731
290	306.446	340	359.282	390	412.117	440	464.953	490	517.788
1	307.503	1	360.338	1	413.174	1	466.009	1	518.845
2	308.559	2	361.395	2	414.230	2	467.066	2	519.902
3	309.616	3	362.452	3	415.287	3	468.123	3	520.958
4	310.673	4	363.508	4	416.344	4	469.179	4	522.015
5	311.730	5	364.565	5	417.401	5	470.236	5	523.072
6	312.786	6	365.622	6	418.457	6	471.293	6	524.128
7	313.843	7	366.679	7	419.514	7	472.350	7	525.185
8	314.900	8	367.735	8	420.571	8	473.406	8	526.242
9	315.956	9	368.792	9	421.627	9	474.463	9	527.298

IX. CAPACITY—LITRES TO LIQUID QUARTS

[Reduction factor: 1 litre = 1.0567104 quarts]

Litres	Liquid quarts	Litres	Liquid quarts	Litres	Liquid quarts	Litres	Liquid quarts	Litres	Liquid quarts
500	528.355	550	581.191	600	634.026	650	686.862	700	739.697
1	529.412	1	582.247	1	635.083	1	687.918	1	740.754
2	530.469	2	583.304	2	636.140	2	688.975	2	741.811
3	531.525	3	584.361	3	637.196	3	690.032	3	742.867
4	532.582	4	585.418	4	638.253	4	691.089	4	743.924
5	533.639	5	586.474	5	639.310	5	692.145	5	744.981
6	534.695	6	587.531	6	640.367	6	693.202	6	746.038
7	535.752	7	588.588	7	641.423	7	694.259	7	747.094
8	536.809	8	589.644	8	642.480	8	695.315	8	748.151
9	537.866	9	590.701	9	643.537	9	696.372	9	749.208
510	538.922	560	591.758	610	644.593	660	697.429	710	750.264
1	539.979	1	592.815	1	645.650	1	698.486	1	751.321
2	541.036	2	593.871	2	646.707	2	699.542	2	752.378
3	542.092	3	594.928	3	647.763	3	700.599	3	753.435
4	543.149	4	595.985	4	648.820	4	701.656	4	754.491
5	544.206	5	597.041	5	649.877	5	702.712	5	755.548
6	545.263	6	598.098	6	650.934	6	703.769	6	756.605
7	546.319	7	599.155	7	651.990	7	704.826	7	757.661
8	547.376	8	600.212	8	653.047	8	705.883	8	758.718
9	548.433	9	601.268	9	654.104	9	706.939	9	759.775
520	549.489	570	602.325	620	655.160	670	707.996	720	760.831
1	550.546	1	603.382	1	656.217	1	709.053	1	761.888
2	551.603	2	604.438	2	657.274	2	710.109	2	762.945
3	552.660	3	605.495	3	658.331	3	711.166	3	764.002
4	553.716	4	606.552	4	659.387	4	712.223	4	765.058
5	554.773	5	607.608	5	660.444	5	713.280	5	766.115
6	555.830	6	608.665	6	661.501	6	714.336	6	767.172
7	556.886	7	609.722	7	662.557	7	715.393	7	768.228
8	557.943	8	610.779	8	663.614	8	716.450	8	769.285
9	559.000	9	611.835	9	664.671	9	717.506	9	770.342
530	560.057	580	612.892	630	665.728	680	718.563	730	771.399
1	561.113	1	613.949	1	666.784	1	719.620	1	772.455
2	562.170	2	615.005	2	667.841	2	720.676	2	773.512
3	563.227	3	616.062	3	668.898	3	721.733	3	774.569
4	564.283	4	617.119	4	669.954	4	722.790	4	775.625
5	565.340	5	618.176	5	671.011	5	723.847	5	776.682
6	566.397	6	619.232	6	672.068	6	724.903	6	777.739
7	567.453	7	620.289	7	673.125	7	725.960	7	778.796
8	568.510	8	621.346	8	674.181	8	727.017	8	779.852
9	569.567	9	622.402	9	675.238	9	728.073	9	780.909
540	570.624	590	623.459	640	676.295	690	729.130	740	781.966
1	571.680	1	624.516	1	677.351	1	730.187	1	783.022
2	572.737	2	625.573	2	678.408	2	731.244	2	784.079
3	573.794	3	626.629	3	679.465	3	732.300	3	785.136
4	574.850	4	627.686	4	680.521	4	733.357	4	786.193
5	575.907	5	628.743	5	681.578	5	734.414	5	787.249
6	576.964	6	629.799	6	682.635	6	735.470	6	788.306
7	578.021	7	630.856	7	683.692	7	736.527	7	789.363
8	579.077	8	631.913	8	684.748	8	737.584	8	790.419
9	580.134	9	632.970	9	685.805	9	738.641	9	791.476

IX. CAPACITY—LITRES TO LIQUID QUARTS

[Reduction factor: 1 litre = 1.0567104 quarts]

Litres	Liquid quarts	Litres	Liquid quarts	Litres	Liquid quarts	Litres	Liquid quarts	Litres	Liquid quarts
750	792.533	800	845.368	850	898.204	900	951.039	950	1,003.875
1	793.590	1	846.425	1	899.261	1	952.096	1	1,004.932
2	794.646	2	847.482	2	900.317	2	953.153	2	1,005.988
3	795.703	3	848.538	3	901.374	3	954.209	3	1,007.045
4	796.760	4	849.595	4	902.431	4	955.266	4	1,008.102
5	797.816	5	850.652	5	903.487	5	956.323	5	1,009.158
6	798.873	6	851.709	6	904.544	6	957.380	6	1,010.215
7	799.930	7	852.765	7	905.601	7	958.436	7	1,011.272
8	800.986	8	853.822	8	906.658	8	959.493	8	1,012.329
9	802.043	9	854.879	9	907.714	9	960.550	9	1,013.385
760	803.100	810	855.935	860	908.771	910	961.606	960	1,014.442
1	804.157	1	856.992	1	909.828	1	962.663	1	1,015.499
2	805.213	2	858.049	2	910.884	2	963.720	2	1,016.555
3	806.270	3	859.106	3	911.941	3	964.777	3	1,017.612
4	807.327	4	860.162	4	912.998	4	965.833	4	1,018.669
5	808.383	5	861.219	5	914.054	5	966.890	5	1,019.726
6	809.440	6	862.276	6	915.111	6	967.947	6	1,020.782
7	810.497	7	863.332	7	916.168	7	969.003	7	1,021.839
8	811.554	8	864.389	8	917.225	8	970.060	8	1,022.896
9	812.610	9	865.446	9	918.281	9	971.117	9	1,023.952
770	813.667	820	866.503	870	919.338	920	972.174	970	1,025.009
1	814.724	1	867.559	1	920.395	1	973.230	1	1,026.066
2	815.780	2	868.616	2	921.451	2	974.287	2	1,027.123
3	816.837	3	869.673	3	922.508	3	975.344	3	1,028.179
4	817.894	4	870.729	4	923.565	4	976.400	4	1,029.236
5	818.951	5	871.786	5	924.622	5	977.457	5	1,030.293
6	820.007	6	872.843	6	925.678	6	978.514	6	1,031.349
7	821.064	7	873.900	7	926.735	7	979.571	7	1,032.406
8	822.121	8	874.956	8	927.792	8	980.627	8	1,033.463
9	823.177	9	876.013	9	928.848	9	981.684	9	1,034.519
780	824.234	830	877.070	880	929.905	930	982.741	980	1,035.576
1	825.291	1	878.126	1	930.962	1	983.797	1	1,036.633
2	826.348	2	879.183	2	932.019	2	984.854	2	1,037.690
3	827.404	3	880.240	3	933.075	3	985.911	3	1,038.746
4	828.461	4	881.296	4	934.132	4	986.968	4	1,039.803
5	829.518	5	882.353	5	935.189	5	988.024	5	1,040.860
6	830.574	6	883.410	6	936.245	6	989.081	6	1,041.916
7	831.631	7	884.467	7	937.302	7	990.138	7	1,042.973
8	832.688	8	885.523	8	938.359	8	991.194	8	1,044.030
9	833.745	9	886.580	9	939.416	9	992.251	9	1,045.087
790	834.801	840	887.637	890	940.472	940	993.308	990	1,046.143
1	835.858	1	888.693	1	941.529	1	994.364	1	1,047.200
2	836.915	2	889.750	2	942.586	2	995.421	2	1,048.257
3	837.971	3	890.807	3	943.642	3	996.478	3	1,049.313
4	839.028	4	891.864	4	944.699	4	997.535	4	1,050.370
5	840.085	5	892.920	5	945.756	5	998.591	5	1,051.427
6	841.141	6	893.977	6	946.813	6	999.648	6	1,052.484
7	842.198	7	895.034	7	947.869	7	1,000.705	7	1,053.540
8	843.255	8	896.090	8	948.926	8	1,001.761	8	1,054.597
9	844.312	9	897.147	9	949.983	9	1,002.818	9	1,055.654

IX. CAPACITY—LIQUID QUARTS TO LITRES

[Reduction factor: 1 liquid quart = 0.94633307 litre]

Liquid quarts	Litres	Liquid quarts	Litres	Liquid quarts	Litres	Liquid quarts	Litres	Liquid quarts	Litres
0		50	47.3167	100	94.633	150	141.950	200	189.267
1	0.9463	1	48.2630	1	95.580	1	142.896	1	190.213
2	1.8927	2	49.2093	2	96.526	2	143.843	2	191.159
3	2.8390	3	50.1557	3	97.472	3	144.789	3	192.106
4	3.7853	4	51.1020	4	98.419	4	145.735	4	193.052
5	4.7317	5	52.0483	5	99.365	5	146.682	5	193.998
6	5.6780	6	52.9947	6	100.311	6	147.628	6	194.945
7	6.6243	7	53.9410	7	101.258	7	148.574	7	195.891
8	7.5707	8	54.8873	8	102.204	8	149.521	8	196.837
9	8.5170	9	55.8337	9	103.150	9	150.467	9	197.784
10	9.4633	60	56.7800	110	104.097	160	151.413	210	198.730
1	10.4097	1	57.7263	1	105.043	1	152.360	1	199.676
2	11.3560	2	58.6727	2	105.989	2	153.306	2	200.623
3	12.3023	3	59.6190	3	106.936	3	154.252	3	201.569
4	13.2487	4	60.5653	4	107.882	4	155.199	4	202.515
5	14.1950	5	61.5116	5	108.828	5	156.145	5	203.462
6	15.1413	6	62.4580	6	109.775	6	157.091	6	204.408
7	16.0877	7	63.4043	7	110.721	7	158.038	7	205.354
8	17.0340	8	64.3506	8	111.667	8	158.984	8	206.301
9	17.9803	9	65.2970	9	112.614	9	159.930	9	207.247
20	18.9267	70	66.2433	120	113.560	170	160.877	220	208.193
1	19.8730	1	67.1896	1	114.506	1	161.823	1	209.140
2	20.8193	2	68.1360	2	115.453	2	162.769	2	210.086
3	21.7657	3	69.0823	3	116.399	3	163.716	3	211.032
4	22.7120	4	70.0286	4	117.345	4	164.662	4	211.979
5	23.6583	5	70.9750	5	118.292	5	165.608	5	212.925
6	24.6047	6	71.9213	6	119.238	6	166.555	6	213.871
7	25.5510	7	72.8676	7	120.184	7	167.501	7	214.818
8	26.4973	8	73.8140	8	121.131	8	168.447	8	215.764
9	27.4437	9	74.7603	9	122.077	9	169.394	9	216.710
30	28.3900	80	75.7066	130	123.023	180	170.340	230	217.657
1	29.3363	1	76.6530	1	123.970	1	171.286	1	218.603
2	30.2827	2	77.5993	2	124.916	2	172.233	2	219.549
3	31.2290	3	78.5456	3	125.862	3	173.179	3	220.496
4	32.1753	4	79.4920	4	126.809	4	174.125	4	221.442
5	33.1217	5	80.4383	5	127.755	5	175.072	5	222.388
6	34.0680	6	81.3846	6	128.701	6	176.018	6	223.335
7	35.0143	7	82.3310	7	129.648	7	176.964	7	224.281
8	35.9607	8	83.2773	8	130.594	8	177.911	8	225.227
9	36.9070	9	84.2236	9	131.540	9	178.857	9	226.174
40	37.8533	90	85.1700	140	132.487	190	179.803	240	227.120
1	38.7997	1	86.1163	1	133.433	1	180.750	1	228.066
2	39.7460	2	87.0626	2	134.379	2	181.696	2	229.013
3	40.6923	3	88.0090	3	135.326	3	182.642	3	229.959
4	41.6387	4	88.9553	4	136.272	4	183.589	4	230.905
5	42.5850	5	89.9016	5	137.218	5	184.535	5	231.852
6	43.5313	6	90.8480	6	138.165	6	185.481	6	232.798
7	44.4777	7	91.7943	7	139.111	7	186.428	7	233.744
8	45.4240	8	92.7406	8	140.057	8	187.374	8	234.691
9	46.3703	9	93.6870	9	141.004	9	188.320	9	235.637

IX. CAPACITY—LIQUID QUARTS TO LITRES

[Reduction factor: 1 liquid quart = 0.94633307 litre]

Liquid quarts	Litres	Liquid quarts	Litres	Liquid quarts	Litres	Liquid quarts	Litres	Liquid quarts	Litres
250	236.583	300	283.900	350	331.217	400	378.533	450	425.850
1	237.530	1	284.846	1	332.163	1	379.480	1	426.796
2	238.476	2	285.793	2	333.109	2	380.426	2	427.743
3	239.422	3	286.739	3	334.056	3	381.372	3	428.689
4	240.369	4	287.685	4	335.002	4	382.319	4	429.635
5	241.315	5	288.632	5	335.948	5	383.265	5	430.582
6	242.261	6	289.578	6	336.895	6	384.211	6	431.528
7	243.208	7	290.524	7	337.841	7	385.158	7	432.474
8	244.154	8	291.471	8	338.787	8	386.104	8	433.421
9	245.100	9	292.417	9	339.734	9	387.050	9	434.367
260	246.047	310	293.363	360	340.680	410	387.997	460	435.313
1	246.993	1	294.310	1	341.626	1	388.943	1	436.260
2	247.939	2	295.256	2	342.573	2	389.889	2	437.206
3	248.886	3	296.202	3	343.519	3	390.836	3	438.152
4	249.832	4	297.149	4	344.465	4	391.782	4	439.099
5	250.778	5	298.095	5	345.412	5	392.728	5	440.045
6	251.725	6	299.041	6	346.358	6	393.675	6	440.991
7	252.671	7	299.988	7	347.304	7	394.621	7	441.938
8	253.617	8	300.934	8	348.251	8	395.567	8	442.884
9	254.564	9	301.880	9	349.197	9	396.514	9	443.830
270	255.510	320	302.827	370	350.143	420	397.460	470	444.777
1	256.456	1	303.773	1	351.090	1	398.406	1	445.723
2	257.403	2	304.719	2	352.036	2	399.353	2	446.669
3	258.349	3	305.666	3	352.982	3	400.299	3	447.616
4	259.295	4	306.612	4	353.929	4	401.245	4	448.562
5	260.242	5	307.558	5	354.875	5	402.192	5	449.508
6	261.188	6	308.505	6	355.821	6	403.138	6	450.455
7	262.134	7	309 451	7	356.768	7	404.084	7	451.401
8	263.081	8	310.397	8	357.714	8	405.031	8	452.347
9	264.027	9	311.344	9	358.660	9	405.977	9	453.294
280	264.973	330	312.290	380	359.607	430	406.923	480	454.240
1	265.920	1	313.236	1	360.553	1	407.870	1	455.186
2	266.866	2	314.183	2	361.499	2	408.816	2	456.133
3	267.812	3	315.129	3	362.446	3	409.762	3	457.079
4	268.759	4	316.075	4	363.392	4	410.709	4	458.025
5	269.705	5	317.022	5	364.338	5	411.655	5	458.972
6	270.651	6	317.968	6	365.285	6	412.601	6	459.918
7	271.598	7	318.914	7	366.231	7	413.548	7	460.864
8	272.544	8	319.861	8	367.177	8	414.494	8	461.811
9	273.490	9	320.807	9	368.124	9	415.440	9	462.757
290	274.437	340	321.753	390	369.070	440	416.387	490	463.703
1	275.383	1	322.700	1	370.016	1	417.333	1	464.650
2	276.329	2	323.646	2	370.963	2	418.279	2	465.596
3	277.276	3	324.592	3	371.909	3	419.226	3	466.542
4	278.222	4	325.539	4	372.855	4	420.172	4	467.489
5	279.168	5	326.485	5	373.802	5	421.118	5	468.435
6	280.115	6	327.431	6	374.748	6	422.065	6	469.381
7	281.061	7	328.378	7	375.694	7	423.011	7	470.328
8	282.007	8	329.324	8	376.641	8	423.957	8	471.274
9	282.954	9	330.270	9	377.587	9	424.904	9	472.220

IX. CAPACITY—LIQUID QUARTS TO LITRES

[Reduction factor: 1 liquid quart = 0.94633307 litre]

Liquid quarts	Litres	Liquid quarts	Litres	Liquid quarts	Litres	Liquid quarts	Litres	Liquid quarts	Litres
500	473.167	550	520.483	600	567.800	650	615.116	700	662.433
1	474.113	1	521.430	1	568.746	1	616.063	1	663.379
2	475.059	2	522.376	2	569.693	2	617.009	2	664.326
3	476.006	3	523.322	3	570.639	3	617.955	3	665.272
4	476.952	4	524.269	4	571.585	4	618.902	4	666.218
5	477.898	5	525.215	5	572.532	5	619.848	5	667.165
6	478.845	6	526.161	6	573.478	6	620.794	6	668.111
7	479.791	7	527.108	7	574.424	7	621.741	7	669.057
8	480.737	8	528.054	8	575.371	8	622.687	8	670.004
9	481.684	9	529.000	9	576.317	9	623.633	9	670.950
510	482.630	560	529.947	610	577.263	660	624.580	710	671.896
1	483.576	1	530.893	1	578.210	1	625.526	1	672.843
2	484.523	2	531.839	2	579.156	2	626.472	2	673.789
3	485.469	3	532.786	3	580.102	3	627.419	3	674.735
4	486.415	4	533.732	4	581.049	4	628.365	4	675.682
5	487.362	5	534.678	5	581.995	5	629.311	5	676.628
6	488.308	6	535.625	6	582.941	6	630.258	6	677.574
7	489.254	7	536.571	7	583.888	7	631.204	7	678.521
8	490.201	8	537.517	8	584.834	8	632.150	8	679.467
9	491.147	9	538.464	9	585.780	9	633.097	9	680.413
520	492.093	570	539.410	620	586.727	670	634.043	720	681.360
1	493.040	1	540.356	1	587.673	1	634.989	1	682.306
2	493.986	2	541.303	2	588.619	2	635.936	2	683.252
3	494.932	3	542.249	3	589.566	3	636.882	3	684.199
4	495.879	4	543.195	4	590.512	4	637.828	4	685.145
5	496.825	5	544.142	5	591.458	5	638.775	5	686.091
6	497.771	6	545.088	6	592.405	6	639.721	6	687.038
7	498.718	7	546.034	7	593.351	7	640.667	7	687.984
8	499.664	8	546.981	8	594.297	8	641.614	8	688.930
9	500.610	9	547.927	9	595.244	9	642.560	9	689.877
530	501.557	580	548.873	630	596.190	680	643.506	730	690.823
1	502.503	1	549.820	1	597.136	1	644.453	1	691.769
2	503.449	2	550.766	2	598.083	2	645.399	2	692.716
3	504.396	3	551.712	3	599.029	3	646.345	3	693.662
4	505.342	4	552.659	4	599.975	4	647.292	4	694.608
5	506.288	5	553.605	5	600.922	5	648.238	5	695.555
6	507.235	6	554.551	6	601.868	6	649.184	6	696.501
7	508.181	7	555.498	7	602.814	7	650.131	7	697.447
8	509.127	8	556.444	8	603.761	8	651.077	8	698.394
9	510.074	9	557.390	9	604.707	9	652.023	9	699.340
540	511.020	590	558.337	640	605.653	690	652.970	740	700.286
1	511.966	1	559.283	1	606.599	1	653.916	1	701.233
2	512.913	2	560.229	2	607.546	2	654.862	2	702.179
3	513.859	3	561.176	3	608.492	3	655.809	3	703.125
4	514.805	4	562.122	4	609.438	4	656.755	4	704.072
5	515.752	5	563.068	5	610.385	5	657.701	5	705.018
6	516.698	6	564.015	6	611.331	6	658.648	6	705.964
7	517.644	7	564.961	7	612.277	7	659.594	7	706.911
8	518.591	8	565.907	8	613.224	8	660.540	8	707.857
9	519.537	9	566.854	9	614.170	9	661.487	9	708.803

IX. CAPACITY—LIQUID QUARTS TO LITRES

[Reduction factor: 1 liquid quart = 0.94633307 litre]

Liquid quarts	Litres	Liquid quarts	Litres	Liquid quarts	Litres	Liquid quarts	Litres	Liquid quarts	Litres
750	709.750	800	757.066	850	804.383	900	851.700	950	899.016
1	710.696	1	758.013	1	805.329	1	852.646	1	899.963
2	711.642	2	758.959	2	806.276	2	853.592	2	900.909
3	712.589	3	759.905	3	807.222	3	854.539	3	901.855
4	713.535	4	760.852	4	808.168	4	855.485	4	902.802
5	714.481	5	761.798	5	809.115	5	856.431	5	903.748
6	715.428	6	762.744	6	810.061	6	857.378	6	904.694
7	716.374	7	763.691	7	811.007	7	858.324	7	905.641
8	717.320	8	764.637	8	811.954	8	859.270	8	906.587
9	718.267	9	765.583	9	812.900	9	860.217	9	907.533
760	719.213	810	766.530	860	813.846	910	861.163	960	908.480
1	720.159	1	767.476	1	814.793	1	862.109	1	909.426
2	721.106	2	768.422	2	815.739	2	863.056	2	910.372
3	722.052	3	769.369	3	816.685	3	864.002	3	911.319
4	722.998	4	770.315	4	817.632	4	864.948	4	912.265
5	723.945	5	771.261	5	818.578	5	865.895	5	913.211
6	724.891	6	772.208	6	819.524	6	866.841	6	914.158
7	725.837	7	773.154	7	820.471	7	867.787	7	915.104
8	726.784	8	774.100	8	821.417	8	868.734	8	916.050
9	727.730	9	775.047	9	822.363	9	869.680	9	916.997
770	728.676	820	775.993	870	823.310	920	870.626	970	917.943
1	729.623	1	776.939	1	824.256	1	871.573	1	918.889
2	730.569	2	777.886	2	825.202	2	872.519	2	919.836
3	731.515	3	778.832	3	826.149	3	873.465	3	920.782
4	732.462	4	779.778	4	827.095	4	874.412	4	921.728
5	733.408	5	780.725	5	828.041	5	875.358	5	922.675
6	734.354	6	781.671	6	828.988	6	876.304	6	923.621
7	735.301	7	782.617	7	829.934	7	877.251	7	924.567
8	736.247	8	783.564	8	830.880	8	878.197	8	925.514
9	737.193	9	784.510	9	831.827	9	879.143	9	926.460
780	738.140	830	785.456	880	832.773	930	880.090	980	927.406
1	739.086	1	786.403	1	833.719	1	881.036	1	928.353
2	740.032	2	787.349	2	834.666	2	881.982	2	929.299
3	740.979	3	788.295	3	835.612	3	882.929	3	930.245
4	741.925	4	789.242	4	836.558	4	883.875	4	931.192
5	742.871	5	790.188	5	837.505	5	884.821	5	932.138
6	743.818	6	791.134	6	838.451	6	885.768	6	933.084
7	744.764	7	792.081	7	839.397	7	886.714	7	934.031
8	745.710	8	793.027	8	840.344	8	887.660	8	934.977
9	746.657	9	793.973	9	841.290	9	888.607	9	935.923
790	747.603	840	794.920	890	842.236	940	889.553	990	936.870
1	748.549	1	795.866	1	843.183	1	890.499	1	937.816
2	749.496	2	796.812	2	844.129	2	891.446	2	938.762
3	750.442	3	797.759	3	845.075	3	892.392	3	939.709
4	751.388	4	798.705	4	846.022	4	893.338	4	940.655
5	752.335	5	799.651	5	846.968	5	894.285	5	941.601
6	753.281	6	800.598	6	847.914	6	895.231	6	942.548
7	754.227	7	801.544	7	848.861	7	896.177	7	943.494
8	755.174	8	802.490	8	849.807	8	897.124	8	944.440
9	756.120	9	803.437	9	850.753	9	898.070	9	945.387

IX. CAPACITY—LITRES TO GALLONS

[Reduction factor: 1 litre = 0.26417760 gallon]

Litres	Gallons	Litres	Gallons	Litres	Gallons	Litres	Gallons	Litres	Gallons
0		50	13.20888	100	26.4178	150	39.6266	200	52.8355
1	0.26418	1	13.47306	1	26.6819	1	39.8908	1	53.0997
2	0.52836	2	13.73724	2	26 9461	2	40.1550	2	53.3639
3	0.79253	3	14.00141	3	27.2103	3	40.4192	3	53.6281
4	1.05671	4	14.26559	4	27.4745	4	40.6834	4	53.8922
5	1.32089	5	14.52977	5	27.7386	5	40.9475	5	54.1564
6	1.58507	6	14.79395	6	28.0028	6	41.2117	6	54.4206
7	1.84924	7	15.05812	7	28.2670	7	41.4759	7	54.6848
8	2.11342	8	15.32230	8	28.5312	8	41.7401	8	54.9489
9	2.37760	9	15.58648	9	28.7954	9	42.0042	9	55.2131
10	2.64178	60	15.85066	110	29.0595	160	42.2684	210	55.4773
1	2.90595	1	16.11483	1	29.3237	1	42.5326	1	55.7415
2	3.17013	2	16.37901	2	29.5879	2	42.7968	2	56.0057
3	3.43431	3	16.64319	3	29.8521	3	43.0609	3	56.2698
4	3.69849	4	16.90737	4	30.1162	4	43.3251	4	56.5340
5	3.96266	5	17.17154	5	30.3804	5	43.5893	5	56.7982
6	4.22684	6	17.43572	6	30.6446	6	43.8535	6	57.0624
7	4.49102	7	17.69990	7	30.9088	7	44.1177	7	57.3265
8	4.75520	8	17.96408	8	31.1730	8	44.3818	8	57.5907
9	5.01937	9	18.22825	9	31.4371	9	44.6460	9	57.8549
20	5.28355	70	18.49243	120	31.7013	170	44.9102	220	58.1191
1	5.54773	1	18.75661	1	31.9655	1	45.1744	1	58.3832
2	5.81191	2	19.02079	2	32.2297	2	45.4385	2	58.6474
3	6.07608	3	19.28496	3	32.4938	3	45.7027	3	58.9116
4	6.34026	4	19.54914	4	32.7580	4	45.9669	4	59.1758
5	6.60444	5	19.81332	5	33.0222	5	46.2311	5	59.4400
6	6.86862	6	20.07750	6	33.2864	6	46.4953	6	59.7041
7	7.13280	7	20.34168	7	33.5506	7	46.7594	7	59.9683
8	7.39697	8	20.60585	8	33.8147	8	47.0236	8	60.2325
9	7.66115	9	20.87003	9	34.0789	9	47.2878	9	60.4967
30	7.92533	80	21.13421	130	34.3431	180	47.5520	230	60.7608
1	8.18951	1	21.39839	1	34.6073	1	47.8161	1	61.0250
2	8.45368	2	21.66256	2	34.8714	2	48.0803	2	61.2892
3	8.71786	3	21.92674	3	35.1356	3	48.3445	3	61.5534
4	8.98204	4	22.19092	4	35.3998	4	48.6087	4	61.8176
5	9.24622	5	22.45510	5	35.6640	5	48.8729	5	62.0817
6	9.51039	6	22.71927	6	35.9282	6	49.1370	6	62.3459
7	9.77457	7	22.98345	7	36.1923	7	49.4012	7	62.6101
8	10.03875	8	23.24763	8	36.4565	8	49.6654	8	62.8743
9	10.30293	9	23.51181	9	36.7207	9	49.9296	9	63.1384
40	10.56710	90	23.77598	140	36.9849	190	50.1937	240	63.4026
1	10.83128	1	24.04016	1	37.2490	1	50.4579	1	63.6668
2	11.09546	2	24.30434	2	37.5132	2	50.7221	2	63.9310
3	11.35964	3	24.56852	3	37.7774	3	50.9863	3	64.1952
4	11.62381	4	24.83269	4	38.0416	4	51.2505	4	64.4593
5	11.88799	5	25.09687	5	38.3058	5	51.5146	5	64.7235
6	12.15217	6	25.36105	6	38.5699	6	51.7788	6	64.9877
7	12.41635	7	25.62523	7	38.8341	7	52.0430	7	65.2519
8	12.68052	8	25.88940	8	39.0983	8	52.3072	8	65.5160
9	12.94470	9	26.15358	9	39.3625	9	52.5713	9	65.7802

IX. CAPACITY—LITRES TO GALLONS

[Reduction factor: 1 litre = 0.26417760 gallon]

Litres	Gallons	Litres	Gallons	Litres	Gallons	Litres	Gallons	Litres	Gallons
250	66.0444	300	79.2533	350	92.4623	400	105.6710	450	118.8799
1	66.3086	1	79.5175	1	92.7263	1	105.9352	1	119.1441
2	66.5728	2	79.7816	2	92.9905	2	106.1994	2	119.4083
3	66.8369	3	80.0458	3	93.2547	3	106.4636	3	119.6725
4	67.1011	4	80.3100	4	93.5189	4	106.7278	4	119.9366
5	67.3653	5	80.5742	5	93.7830	5	106.9919	5	120.2008
6	67.6295	6	80.8383	6	94.0472	6	107.2561	6	120.4650
7	67.8936	7	81.1025	7	94.3114	7	107.5203	7	120.7292
8	68.1578	8	81.3667	8	94.5756	8	107.7845	8	120.9933
9	68.4220	9	81.6309	9	94.8398	9	108.0486	9	121.2575
260	68.6862	310	81.8951	360	95.1039	410	108.3128	460	121.5217
1	68.9504	1	82.1592	1	95.3681	1	108.5770	1	121.7859
2	69.2145	2	82.4234	2	95.6323	2	108.8412	2	122.0501
3	69.4787	3	82.6876	3	95.8965	3	109.1053	3	122.3142
4	69.7429	4	82.9518	4	96.1606	4	109.3695	4	122.5784
5	70.0071	5	83.2159	5	96.4248	5	109.6337	5	122.8426
6	70.2712	6	83.4801	6	96.6890	6	109.8979	6	123.1068
7	70.5354	7	83.7443	7	96.9532	7	110.1621	7	123.3709
8	70.7996	8	84.0085	8	97.2174	8	110.4262	8	123.6351
9	71.0638	9	84.2727	9	97.4815	9	110.6904	9	123.8993
270	71.3280	320	84.5368	370	97.7457	420	110.9546	470	124.1635
1	71.5921	1	84.8010	1	98.0099	1	111.2188	1	124.4276
2	71.8563	2	85.0652	2	98.2741	2	111.4829	2	124.6918
3	72.1205	3	85.3294	3	98.5382	3	111.7471	3	124.9560
4	72.3847	4	85.5935	4	98.8024	4	112.0113	4	125.2202
5	72.6488	5	85.8577	5	99.0666	5	112.2755	5	125.4844
6	72.9130	6	86.1219	6	99.3308	6	112.5397	6	125.7485
7	73.1772	7	86.3861	7	99.5950	7	112.8038	7	126.0127
8	73.4414	8	86.6503	8	99.8591	8	113.0680	8	126.2769
9	73.7056	9	86.9144	9	100.1233	9	113.3322	9	126.5411
280	73.9697	330	87.1786	380	100.3875	430	113.5964	480	126.8052
1	74.2339	1	87.4428	1	100.6517	1	113.8605	1	127.0694
2	74.4981	2	87.7070	2	100.9158	2	114.1247	2	127.3336
3	74.7623	3	87.9711	3	101.1800	3	114.3889	3	127.5978
4	75.0264	4	88.2353	4	101.4442	4	114.6531	4	127.8620
5	75.2906	5	88.4995	5	101.7084	5	114.9173	5	128.1261
6	75.5548	6	88.7637	6	101.9726	6	115.1814	6	128.3903
7	75.8190	7	89.0279	7	102.2367	7	115.4456	7	128.6545
8	76.0831	8	89.2920	8	102.5009	8	115.7098	8	128.9187
9	76.3473	9	89.5562	9	102.7651	9	115.9740	9	129.1828
290	76.6115	340	89.8204	390	103.0293	440	116.2381	490	129.4470
1	76.8757	1	90.0846	1	103.2934	1	116.5023	1	129.7112
2	77.1399	2	90.3487	2	103.5576	2	116.7665	2	129.9754
3	77.4040	3	90.6129	3	103.8218	3	117.0307	3	130.2396
4	77.6682	4	90.8771	4	104.0860	4	117.2949	4	130.5037
5	77.9324	5	91.1413	5	104.3502	5	117.5590	5	130.7679
6	78.1966	6	91.4054	6	104.6143	6	117.8232	6	131.0321
7	78.4607	7	91.6696	7	104.8785	7	118.0874	7	131.2963
8	78.7249	8	91.9338	8	105.1427	8	118.3516	8	131.5604
9	78.9891	9	92.1980	9	105.4069	9	118.6157	9	131.8246

IX. CAPACITY—LITRES TO GALLONS

[Reduction factor: 1 litre = 0.26417760 gallon]

Litres	Gallons	Litres	Gallons	Litres	Gallons	Litres	Gallons	Litres	Gallons
500	132.0888	550	145.2977	600	158.5066	650	171.7154	700	184.9243
1	132.3530	1	145.5619	1	158.7707	1	171.9796	1	185.1885
2	132.6172	2	145.8260	2	159.0349	2	172.2438	2	185.4527
3	132.8813	3	146.0902	3	159.2991	3	172.5080	3	185.7169
4	133.1455	4	146.3544	4	159.5633	4	172.7722	4	185.9810
5	133.4097	5	146.6186	5	159.8274	5	173.0363	5	186.2452
6	133.6739	6	146.8827	6	160.0916	6	173.3005	6	186.5094
7	133.9380	7	147.1469	7	160.3558	7	173.5647	7	186.7736
8	134.2022	8	147.4111	8	160.6200	8	173.8289	8	187.0377
9	134.4664	9	147.6753	9	160.8842	9	174.0930	9	187.3019
510	134.7306	560	147.9395	610	161.1483	660	174.3572	710	187.5661
1	134.9948	1	148.2036	1	161.4125	1	174.6214	1	187.8303
2	135.2589	2	148.4678	2	161.6767	2	174.8856	2	188.0945
3	135.5231	3	148.7320	3	161.9409	3	175.1497	3	188.3586
4	135.7873	4	148.9962	4	162.2050	4	175.4139	4	188.6228
5	136.0515	5	149.2603	5	162.4692	5	175.6781	5	188.8870
6	136.3156	6	149.5245	6	162.7334	6	175.9423	6	189.1512
7	136.5798	7	149.7887	7	162.9976	7	176.2065	7	189.4153
8	136.8440	8	150.0529	8	163.2618	8	176.4706	8	189.6795
9	137.1082	9	150.3171	9	163.5259	9	176.7348	9	189.9437
520	137.3724	570	150.5812	620	163.7901	670	176.9990	720	190.2079
1	137.6365	1	150.8454	1	164.0543	1	177.2632	1	190.4720
2	137.9007	2	151.1096	2	164.3185	2	177.5273	2	190.7362
3	138.1649	3	151.3738	3	164.5826	3	177.7915	3	191.0004
4	138.4291	4	151.6379	4	164.8468	4	178.0557	4	191.2646
5	138.6932	5	151.9021	5	165.1110	5	178.3199	5	191.5288
6	138.9574	6	152.1663	6	165.3752	6	178.5841	6	191.7929
7	139.2216	7	152.4305	7	165.6394	7	178.8482	7	192.0571
8	139.4858	8	152.6947	8	165.9035	8	179.1124	8	192.3213
9	139.7500	9	152.9588	9	166.1677	9	179.3766	9	192.5855
530	140.0141	580	153.2230	630	166.4319	680	179.6408	730	192.8496
1	140.2783	1	153.4872	1	166.6961	1	179.9049	1	193.1138
2	140.5425	2	153.7514	2	166.9602	2	180.1691	2	193.3780
3	140.8067	3	154.0155	3	167.2244	3	180.4333	3	193.6422
4	141.0708	4	154.2797	4	167.4886	4	180.6975	4	193.9064
5	141.3350	5	154.5439	5	167.7528	5	180.9617	5	194.1705
6	141.5992	6	154.8081	6	168.0170	6	181.2258	6	194.4347
7	141.8634	7	155.0723	7	168.2811	7	181.4900	7	194.6989
8	142.1275	8	155.3364	8	168.5453	8	181.7542	8	194.9631
9	142.3917	9	155.6006	9	168.8095	9	182.0184	9	195.2272
540	142.6559	590	155.8648	640	169.0737	690	182.2825	740	195.4914
1	142.9201	1	156.1290	1	169.3378	1	182.5467	1	195.7556
2	143.1843	2	156.3931	2	169.6020	2	182.8109	2	196.0198
3	143.4484	3	156.6573	3	169.8662	3	183.0751	3	196.2840
4	143.7126	4	156.9215	4	170.1304	4	183.3393	4	196.5481
5	143.9768	5	157.1857	5	170.3946	5	183.6034	5	196.8123
6	144.2410	6	157.4498	6	170.6587	6	183.8676	6	197.0765
7	144.5051	7	157.7140	7	170.9229	7	184.1318	7	197.3407
8	144.7693	8	157.9782	8	171.1871	8	184.3960	8	197.6048
9	145.0335	9	158.2424	9	171.4513	9	184.6601	9	197.8690

IX. CAPACITY—LITRES TO GALLONS

[Reduction factor: 1 litre = 0.26417760 gallon]

Litres	Gallons	Litres	Gallons	Litres	Gallons	Litres	Gallons	Litres	Gallons
750	198.1332	**800**	211.3421	**850**	224.5510	**900**	237.7598	**950**	250.9687
1	198.3974	1	211.6063	1	224.8151	1	238.0240	1	251.2329
2	198.6616	2	211.8704	2	225.0793	2	238.2882	2	251.4971
3	198.9257	3	212.1346	3	225.3435	3	238.5524	3	251.7613
4	199.1899	4	212.3988	4	225.6077	4	238.8166	4	252.0254
5	199.4541	5	212.6630	5	225.8718	5	239.0807	5	252.2896
6	199.7183	6	212.9271	6	226.1360	6	239.3449	6	252.5538
7	199.9824	7	213.1913	7	226.4002	7	239.6091	7	252.8180
8	200.2466	8	213.4555	8	226.6644	8	239.8733	8	253.0821
9	200.5108	9	213.7197	9	226.9286	9	240.1374	9	253.3463
760	200.7750	**810**	213.9839	**860**	227.1927	**910**	240.4016	**960**	253.6105
1	201.0392	1	214.2480	1	227.4569	1	240.6658	1	253.8747
2	201.3033	2	214.5122	2	227.7211	2	240.9300	2	254.1389
3	201.5675	3	214.7764	3	227.9853	3	241.1941	3	254.4030
4	201.8317	4	215.0406	4	228.2494	4	241.4583	4	254.6672
5	202.0959	5	215.3047	5	228.5136	5	241.7225	5	254.9314
6	202.3600	6	215.5689	6	228.7778	6	241.9867	6	255.1956
7	202.6242	7	215.8331	7	229.0420	7	242.2509	7	255.4597
8	202.8884	8	216.0973	8	229.3062	8	242.5150	8	255.7239
9	203.1526	9	216.3615	9	229.5703	9	242.7792	9	255.9881
770	203.4168	**820**	216.6256	**870**	229.8345	**920**	243.0434	**970**	256.2523
1	203.6809	1	216.8898	1	230.0987	1	243.3076	1	256.5164
2	203.9451	2	217.1540	2	230.3629	2	243.5717	2	256.7806
3	204.2093	3	217.4182	3	230.6270	3	243.8359	3	257.0448
4	204.4735	4	217.6823	4	230.8912	4	244.1001	4	257.3090
5	204.7376	5	217.9465	5	231.1554	5	244.3643	5	257.5732
6	205.0018	6	218.2107	6	231.4196	6	244.6285	6	257.8373
7	205.2660	7	218.4749	7	231.6838	7	244.8926	7	258.1015
8	205.5302	8	218.7391	8	231.9479	8	245.1568	8	258.3657
9	205.7944	9	219.0032	9	232.2121	9	245.4210	9	258.6299
780	206.0585	**830**	219.2674	**880**	232.4763	**930**	245.6852	**980**	258.8940
1	206.3227	1	219.5316	1	232.7405	1	245.9493	1	259.1582
2	206.5869	2	219.7958	2	233.0046	2	246.2135	2	259.4224
3	206.8511	3	220.0599	3	233.2688	3	246.4777	3	259.6866
4	207.1152	4	220.3241	4	233.5330	4	246.7419	4	259.9508
5	207.3794	5	220.5883	5	233.7972	5	247.0061	5	260.2149
6	207.6436	6	220.8525	6	234.0614	6	247.2702	6	260.4791
7	207.9078	7	221.1167	7	234.3255	7	247.5344	7	260.7433
8	208.1719	8	221.3808	8	234.5897	8	247.7986	8	261.0075
9	208.4361	9	221.6450	9	234.8539	9	248.0628	9	261.2716
790	208.7003	**840**	221.9092	**890**	235.1181	**940**	248.3269	**990**	261.5358
1	208.9645	1	222.1734	1	235.3822	1	248.5911	1	261.8000
2	209.2287	2	222.4375	2	235.6464	2	248.8553	2	262.0642
3	209.4928	3	222.7017	3	235.9106	3	249.1195	3	262.3284
4	209.7570	4	222.9659	4	236.1748	4	249.3837	4	262.5925
5	210.0212	5	223.2301	5	236.4390	5	249.6478	5	262.8567
6	210.2854	6	223.4942	6	236.7031	6	249.9120	6	263.1209
7	210.5495	7	223.7584	7	236.9673	7	250.1762	7	263.3851
8	210.8137	8	224.0226	8	237.2315	8	250.4404	8	263.6492
9	211.0779	9	224.2868	9	237.4957	9	250.7045	9	263.9134

133

IX. CAPACITY—GALLONS TO LITRES

[Reduction factor: 1 gallon = 3.7853323 litres]

Gallons	Litres	Gallons	Litres	Gallons	Litres	Gallons	Litres	Gallons	Litres
0		50	189.2666	100	378.533	150	567.800	200	757.066
1	3.7853	1	193.0519	1	382.319	1	571.585	1	760.852
2	7.5707	2	196.8373	2	386.104	2	575.371	2	764.637
3	11.3560	3	200.6226	3	389.889	3	579.156	3	768.422
4	15.1413	4	204.4079	4	393.675	4	582.941	4	772.208
5	18.9267	5	208.1933	5	397.460	5	586.727	5	775.993
6	22.7120	6	211.9786	6	401.245	6	590.512	6	779.778
7	26.4973	7	215.7639	7	405.031	7	594.297	7	783.564
8	30.2827	8	219.5493	8	408.816	8	598.083	8	787.349
9	34.0680	9	223.3346	9	412.601	9	601.868	9	791.134
10	37.8533	60	227.1199	110	416.387	160	605.653	210	794.920
1	41.6387	1	230.9053	1	420.172	1	609.438	1	798.705
2	45.4240	2	234.6906	2	423.957	2	613.224	2	802.490
3	49.2093	3	238.4759	3	427.743	3	617.009	3	806.276
4	52.9947	4	242.2613	4	431.528	4	620.794	4	810.061
5	56.7800	5	246.0466	5	435.313	5	624.580	5	813.846
6	60.5653	6	249.8319	6	439.099	6	628.365	6	817.632
7	64.3506	7	253.6173	7	442.884	7	632.150	7	821.417
8	68.1360	8	257.4026	8	446.669	8	635.936	8	825.202
9	71.9213	9	261.1879	9	450.455	9	639.721	9	828.988
20	75.7066	70	264.9733	120	454.240	170	643.506	220	832.773
1	79.4920	1	268.7586	1	458.025	1	647.292	1	836.558
2	83.2773	2	272.5439	2	461.811	2	651.077	2	840.344
3	87.0626	3	276.3293	3	465.596	3	654.862	3	844.129
4	90.8480	4	280.1146	4	469.381	4	658.648	4	847.914
5	94.6333	5	283.8999	5	473.167	5	662.433	5	851.700
6	98.4186	6	287.6853	6	476.952	6	666.218	6	855.485
7	102.2040	7	291.4706	7	480.737	7	670.004	7	859.270
8	105.9893	8	295.2559	8	484.523	8	673.789	8	863.056
9	109.7746	9	299.0413	9	488.308	9	677.574	9	866.841
30	113.5600	80	302.8266	130	492.093	180	681.360	230	870.626
1	117.3453	1	306.6119	1	495.879	1	685.145	1	874.412
2	121.1306	2	310.3972	2	499.664	2	688.930	2	878.197
3	124.9160	3	314.1826	3	503.449	3	692.716	3	881.982
4	128.7013	4	317.9679	4	507.235	4	696.501	4	885.768
5	132.4866	5	321.7532	5	511.020	5	700.286	5	889.553
6	136.2720	6	325.5386	6	514.805	6	704.072	6	893.338
7	140.0573	7	329.3239	7	518.591	7	707.857	7	897.124
8	143.8426	8	333.1092	8	522.376	8	711.642	8	900.909
9	147.6280	9	336.8946	9	526.161	9	715.428	9	904.694
40	151.4133	90	340.6799	140	529.947	190	719.213	240	908.480
1	155.1986	1	344.4652	1	533.732	1	722.998	1	912.265
2	158.9840	2	348.2506	2	537.517	2	726.784	2	916.050
3	162.7693	3	352.0359	3	541.303	3	730.569	3	919.836
4	166.5546	4	355.8212	4	545.088	4	734.354	4	923.621
5	170.3400	5	359.6066	5	548.873	5	738.140	5	927.406
6	174.1253	6	363.3919	6	552.659	6	741.925	6	931.192
7	177.9106	7	367.1772	7	556.444	7	745.710	7	934.977
8	181.6960	8	370.9626	8	560.229	8	749.496	8	938.762
9	185.4813	9	374.7479	9	564.015	9	753.281	9	942.548

IX. CAPACITY—GALLONS TO LITRES

[Reduction factor: 1 gallon = 3.7853323 litres]

Gallons	Litres	Gallons	Litres	Gallons	Litres	Gallons	Litres	Gallons	Litres
250	946.333	300	1,135.600	350	1,324.866	400	1,514.133	450	1,703.400
1	950.118	1	1,139.385	1	1,328.652	1	1,517.918	1	1,707.185
2	953.904	2	1,143.170	2	1,332.437	2	1,521.704	2	1,710.970
3	957.689	3	1,146.956	3	1,336.222	3	1,525.489	3	1,714.756
4	961.474	4	1,150.741	4	1,340.008	4	1,529.274	4	1,718.541
5	965.260	5	1,154.526	5	1,343.793	5	1,533.060	5	1,722.326
6	969.045	6	1,158.312	6	1,347.578	6	1,536.845	6	1,726.112
7	972.830	7	1,162.097	7	1,351.364	7	1,540.630	7	1,729.897
8	976.616	8	1,165.882	8	1,355.149	8	1,544.416	8	1,733.682
9	980.401	9	1,169.668	9	1,358.934	9	1,548.201	9	1,737.468
260	984.186	310	1,173.453	360	1,362.720	410	1,551.986	460	1,741.253
1	987.972	1	1,177.238	1	1,366.525	1	1,555.772	1	1,745.038
2	991.757	2	1,181.024	2	1,370.290	2	1,559.557	2	1,748.824
3	995.542	3	1,184.809	3	1,374.076	3	1,563.342	3	1,752.609
4	999.328	4	1,188.594	4	1,377.851	4	1,567.128	4	1,756.394
5	1,003.113	5	1,192.380	5	1,381.646	5	1,570.913	5	1,760.180
6	1,006.898	6	1,196.165	6	1,385.432	6	1,574.698	6	1,763.965
7	1,010.684	7	1,199.950	7	1,389.217	7	1,578.484	7	1,767.750
8	1,014.469	8	1,203.736	8	1,393.002	8	1,582.269	8	1,771.536
9	1,018.254	9	1,207.521	9	1,396.788	9	1,586.054	9	1,775.321
270	1,022.040	320	1,211.306	370	1,400.573	420	1,589.840	470	1,779.106
1	1,025.825	1	1,215.092	1	1,404.358	1	1,593.625	1	1,782.892
2	1,029.610	2	1,218.877	2	1,408.144	2	1,597.410	2	1,786.677
3	1,033.396	3	1,222.662	3	1,411.929	3	1,601.196	3	1,790.462
4	1,037.181	4	1,226.448	4	1,415.714	4	1,604.981	4	1,794.248
5	1,040.966	5	1,230.233	5	1,419.500	5	1,608.766	5	1,798.033
6	1,044.752	6	1,234.018	6	1,423.285	6	1,612.552	6	1,801.818
7	1,048.537	7	1,237.804	7	1,427.070	7	1,616.337	7	1,805.604
8	1,052.322	8	1,241.589	8	1,430.856	8	1,620.122	8	1,809.389
9	1,056.108	9	1,245.374	9	1,434.641	9	1,623.908	9	1,813.174
280	1,059.893	330	1,249.160	380	1,438.426	430	1,627.693	480	1,816.960
1	1,063.678	1	1,252.945	1	1,442.212	1	1,631.478	1	1,820.745
2	1,067.464	2	1,256.730	2	1,445.997	2	1,635.264	2	1,824.530
3	1,071.249	3	1,260.516	3	1,449.782	3	1,639.049	3	1,828.316
4	1,075.034	4	1,264.301	4	1,453.568	4	1,642.834	4	1,832.101
5	1,078.820	5	1,268.086	5	1,457.353	5	1,646.620	5	1,835.886
6	1,082.605	6	1,271.872	6	1,461.138	6	1,650.405	6	1,839.671
7	1,086.390	7	1,275.657	7	1,464.924	7	1,654.190	7	1,843.457
8	1,090.176	8	1,279.442	8	1,468.709	8	1,657.976	8	1,847.242
9	1,093.961	9	1,283.228	9	1,472.494	9	1,661.761	9	1,851.027
290	1,097.746	340	1,287.013	390	1,476.280	440	1,665.546	490	1,854.813
1	1,101.532	1	1,290.798	1	1,480.065	1	1,669.332	1	1,858.598
2	1,105.317	2	1,294.584	2	1,483.850	2	1,673.117	2	1,862.383
3	1,109.102	3	1,298.369	3	1,487.636	3	1,676.902	3	1,866.169
4	1,112.888	4	1,302.154	4	1,491.421	4	1,680.688	4	1,869.954
5	1,116.673	5	1,305.940	5	1,495.206	5	1,684.473	5	1,873.739
6	1,120.458	6	1,309.725	6	1,498.992	6	1,688.258	6	1,877.525
7	1,124.244	7	1,313.510	7	1,502.777	7	1,692.044	7	1,881.310
8	1,128.029	8	1,317.296	8	1,506.562	8	1,695.829	8	1,885.095
9	1,131.814	9	1,321.081	9	1,510.348	9	1,699.614	9	1,888.881

IX. CAPACITY—GALLONS TO LITRES

[Reduction factor: 1 gallon = 3.7853323 litres]

Gallons	Litres	Gallons	Litres	Gallons	Litres	Gallons	Litres	Gallons	Litres
500	1,892.666	550	2,081.933	600	2,271.199	650	2,460.466	700	2,649.733
1	1,896.451	1	2,085.718	1	2,274.985	1	2,464.251	1	2,653.518
2	1,900.237	2	2,089.503	2	2,278.770	2	2,468.037	2	2,657.303
3	1,904.022	3	2,093.289	3	2,282.555	3	2,471.822	3	2,661.089
4	1,907.807	4	2,097.074	4	2,286.341	4	2,475.607	4	2,664.874
5	1,911.593	5	2,100.859	5	2,290.126	5	2,479.393	5	2,668.659
6	1,915.378	6	2,104.645	6	2,293.911	6	2,483.178	6	2,672.445
7	1,919.163	7	2,108.430	7	2,297.697	7	2,486.963	7	2,676.230
8	1,922.949	8	2,112.215	8	2,301.482	8	2,490.749	8	2,680.015
9	1,926.734	9	2,116.001	9	2,305.267	9	2,494.534	9	2,683.801
510	1,930.519	560	2,119.786	610	2,309.053	660	2,498.319	710	2,687.586
1	1,934.305	1	2,123.571	1	2,312.838	1	2,502.105	1	2,691.371
2	1,938.090	2	2,127.357	2	2,316.623	2	2,505.890	2	2,695.157
3	1,941.875	3	2,131.142	3	2,320.409	3	2,509.675	3	2,698.942
4	1,945.661	4	2,134.927	4	2,324.194	4	2,513.461	4	2,702.727
5	1,949.446	5	2,138.713	5	2,327.979	5	2,517.246	5	2,706.513
6	1,953.231	6	2,142.498	6	2,331.765	6	2,521.031	6	2,710.298
7	1,957.017	7	2,146.283	7	2,335.550	7	2,524.817	7	2,714.083
8	1,960.802	8	2,150.069	8	2,339.335	8	2,528.602	8	2,717.869
9	1,964.587	9	2,153.854	9	2,343.121	9	2,532.387	9	2,721.654
520	1,968.373	570	2,157.639	620	2,346.906	670	2,536.173	720	2,725.439
1	1,972.158	1	2,161.425	1	2,350.691	1	2,539.958	1	2,729.225
2	1,975.943	2	2,165.210	2	2,354.477	2	2,543.743	2	2,733.010
3	1,979.729	3	2,168.995	3	2,358.262	3	2,547.529	3	2,736.795
4	1,983.514	4	2,172.781	4	2,362.047	4	2,551.314	4	2,740.581
5	1,987.299	5	2,176.566	5	2,365.833	5	2,555.099	5	2,744.366
6	1,991.085	6	2,180.351	6	2,369.618	6	2,558.885	6	2,748.151
7	1,994.870	7	2,184.137	7	2,373.403	7	2,562.670	7	2,751.937
8	1,998.655	8	2,187.922	8	2,377.189	8	2,566.455	8	2,755.722
9	2,002.441	9	2,191.707	9	2,380.974	9	2,570.241	9	2,759.507
530	2,006.226	580	2,195.493	630	2,384.759	680	2,574.026	730	2,763.293
1	2,010.011	1	2,199.278	1	2,388.545	1	2,577.811	1	2,767.078
2	2,013.797	2	2,203.063	2	2,392.330	2	2,581.597	2	2,770.863
3	2,017.582	3	2,206.849	3	2,396.115	3	2,585.382	3	2,774.649
4	2,021.367	4	2,210.634	4	2,399.901	4	2,589.167	4	2,778.434
5	2,025.153	5	2,214.419	5	2,403.686	5	2,592.953	5	2,782.219
6	2,028.938	6	2,218.205	6	2,407.471	6	2,596.738	6	2,786.005
7	2,032.723	7	2,221.990	7	2,411.257	7	2,600.523	7	2,789.790
8	2,036.509	8	2,225.775	8	2,415.042	8	2,604.309	8	2,793.575
9	2,040.294	9	2,229.561	9	2,418.827	9	2,608.094	9	2,797.361
540	2,044.079	590	2,233.346	640	2,422.613	690	2,611.879	740	2,801.146
1	2,047.865	1	2,237.131	1	2,426.398	1	2,615.665	1	2,804.931
2	2,051.650	2	2,240.917	2	2,430.183	2	2,619.450	2	2,808.717
3	2,055.435	3	2,244.702	3	2,433.969	3	2,623.235	3	2,812.502
4	2,059.221	4	2,248.487	4	2,437.754	4	2,627.021	4	2,816.287
5	2,063.006	5	2,252.273	5	2,441.539	5	2,630.806	5	2,820.073
6	2,066.791	6	2,256.058	6	2,445.325	6	2,634.591	6	2,823.858
7	2,070.577	7	2,259.843	7	2,449.110	7	2,638.377	7	2,827.643
8	2,074.362	8	2,263.629	8	2,452.895	8	2,642.162	8	2,831.429
9	2,078.147	9	2,267.414	9	2,456.681	9	2,645.947	9	2,835.214

IX. CAPACITY—GALLONS TO LITRES

[Reduction factor: 1 gallon = 3.7853323 litres]

Gallons	Litres	Gallons	Litres	Gallons	Litres	Gallons	Litres	Gallons	Litres
750	2,838.999	800	3,028.266	850	3,217.532	900	3,406.799	950	3,596.066
1	2,842.785	1	3,032.051	1	3,221.318	1	3,410.584	1	3,599.851
2	2,846.570	2	3,035.837	2	3,225.103	2	3,414.370	2	3,603.636
3	2,850.355	3	3,039.622	3	3,228.888	3	3,418.155	3	3,607.422
4	2,854.141	4	3,043.407	4	3,232.674	4	3,421.940	4	3,611.207
5	2,857.926	5	3,047.193	5	3,236.459	5	3,425.726	5	3,614.992
6	2,861.711	6	3,050.978	6	3,240.244	6	3,429.511	6	3,618.778
7	2,865.497	7	3,054.763	7	3,244.030	7	3,433.296	7	3,622.563
8	2,869.282	8	3,058.548	8	3,247.815	8	3,437.082	8	3,626.348
9	2,873.067	9	3,062.334	9	3,251.600	9	3,440.867	9	3,630.134
760	2,876.853	810	3,066.119	860	3,255.386	910	3,444.652	960	3,633.919
1	2,880.638	1	3,069.904	1	3,259.171	1	3,448.438	1	3,637.704
2	2,884.423	2	3,073.690	2	3,262.956	2	3,452.223	2	3,641.490
3	2,888.209	3	3,077.475	3	3,266.742	3	3,456.008	3	3,645.275
4	2,891.994	4	3,081.260	4	3,270.527	4	3,459.794	4	3,649.060
5	2,895.779	5	3,085.046	5	3,274.312	5	3,463.579	5	3,652.846
6	2,899.565	6	3,088.831	6	3,278.098	6	3,467.364	6	3,656.631
7	2,903.350	7	3,092.616	7	3,281.883	7	3,471.150	7	3,660.416
8	2,907.135	8	3,096.402	8	3,285.668	8	3,474.935	8	3,664.202
9	2,910.921	9	3,100.187	9	3,289.454	9	3,478.720	9	3,667.987
770	2,914.706	820	3,103.972	870	3,293.239	920	3,482.506	970	3,671.772
1	2,918.491	1	3,107.758	1	3,297.024	1	3,486.291	1	3,675.558
2	2,922.277	2	3,111.543	2	3,300.810	2	3,490.076	2	3,679.343
3	2,926.062	3	3,115.328	3	3,304.595	3	3,493.862	3	3,683.128
4	2,929.847	4	3,119.114	4	3,308.380	4	3,497.647	4	3,686.914
5	2,933.633	5	3,122.899	5	3,312.166	5	3,501.432	5	3,690.699
6	2,937.418	6	3,126.684	6	3,315.951	6	3,505.218	6	3,694.484
7	2,941.203	7	3,130.470	7	3,319.736	7	3,509.003	7	3,698.270
8	2,944.989	8	3,134.255	8	3,323.522	8	3,512.788	8	3,702.055
9	2,948.774	9	3,138.040	9	3,327.307	9	3,516.574	9	3,705.840
780	2,952.559	830	3,141.826	880	3,331.092	930	3,520.359	980	3,709.626
1	2,956.345	1	3,145.611	1	3,334.878	1	3,524.144	1	3,713.411
2	2,960.130	2	3,149.396	2	3,338.663	2	3,527.930	2	3,717.196
3	2.963.915	3	3,153.182	3	3,342.448	3	3,531.715	3	3,720.982
4	2,967.701	4	3,156.967	4	3,346.234	4	3,535.500	4	3,724.767
5	2,971.486	5	3,160.752	5	3,350.019	5	3,539.286	5	3,728.552
6	2,975.271	6	3,164.538	6	3,353.804	6	3,543.071	6	3,732.338
7	2,979.057	7	3,168.323	7	3,357.590	7	3,546.856	7	3,736.123
8	2,982.842	8	3,172.108	8	3,361.375	8	3,550.642	8	3,739.908
9	2,986.627	9	3,175.894	9	3,365.160	9	3,554.427	9	3,743.694
790	2,990.413	840	3,179.679	890	3,368.946	940	3,558.212	990	3,747.479
1	2,994.198	1	3,183.464	1	3,372.731	1	3,561.998	1	3,751.264
2	2,997.983	2	3,187.250	2	3,376.516	2	3,565.783	2	3,755.050
3	3,001.769	3	3,191.035	3	3,380.302	3	3,569.568	3	3,758.835
4	3,005.554	4	3,194.820	4	3,384.087	4	3,573.354	4	3,762.620
5	3,009.339	5	3,198.606	5	3,387.872	5	3,577.139	5	3,766.406
6	3,013.125	6	3,202.391	6	3,391.658	6	3,580.924	6	3,770.191
7	3,016.910	7	3,206.176	7	3,395.443	7	3,584.710	7	3,773.976
8	3,020.695	8	3,209.962	8	3,399.228	8	3,588.495	8	3,777.762
9	3,024.481	9	3,213.747	9	3,403.014	9	3,592.280	9	3,781.547

IX. CAPACITY—HECTOLITRES TO BUSHELS

[Reduction factor: 1 hectolitre = 2.8378189 bushels]

Hecto-litres	Bush-els	Hecto-litres	Bush-els	Hecto-litres	Bush-els	Hecto-litres	Bush-els	Hecto-litres	Bush-els
0		50	141.8909	100	283.782	150	425.673	200	567.564
1	2.8378	1	144.7288	1	286.620	1	428.511	1	570.402
2	5.6756	2	147.5666	2	289.458	2	431.348	2	573.239
3	8.5135	3	150.4044	3	292.295	3	434.186	3	576.077
4	11.3513	4	153.2422	4	295.133	4	437.024	4	578.915
5	14.1891	5	156.0800	5	297.971	5	439.862	5	581.753
6	17.0269	6	158.9179	6	300.809	6	442.700	6	584.591
7	19.8647	7	161.7557	7	303.647	7	445.538	7	587.429
8	22.7026	8	164.5935	8	306.484	8	448.375	8	590.266
9	25.5404	9	167.4313	9	309.322	9	451.213	9	593.104
10	28.3782	60	170.2691	110	312.160	160	454.051	210	595.942
1	31.2160	1	173.1070	1	314.998	1	456.889	1	598.780
2	34.0538	2	175.9448	2	317.836	2	459.727	2	601.618
3	36.8916	3	178.7826	3	320.674	3	462.564	3	604.455
4	39.7295	4	181.6204	4	323.511	4	465.402	4	607.293
5	42.5673	5	184.4582	5	326.349	5	468.240	5	610.131
6	45.4051	6	187.2960	6	329.187	6	471.078	6	612.969
7	48.2429	7	190.1339	7	332.025	7	473.916	7	615.807
8	51.0807	8	192.9717	8	334.863	8	476.754	8	618.645
9	53.9186	9	195.8095	9	337.700	9	479.591	9	621.482
20	56.7564	70	198.6473	120	340.538	170	482.429	220	624.320
1	59.5942	1	201.4851	1	343.376	1	485.267	1	627.158
2	62.4320	2	204.3230	2	346.214	2	488.105	2	629.996
3	65.2698	3	207.1608	3	349.052	3	490.943	3	632.834
4	68.1077	4	209.9986	4	351.890	4	493.780	4	635.671
5	70.9455	5	212.8364	5	354.727	5	496.618	5	638.509
6	73.7833	6	215.6742	6	357.565	6	499.456	6	641.347
7	76.6211	7	218.5121	7	360.403	7	502.294	7	644.185
8	79.4589	8	221.3499	8	363.241	8	505.132	8	647.023
9	82.2967	9	224.1877	9	366.079	9	507.970	9	649.861
30	85.1346	80	227.0255	130	368.916	180	510.807	230	652.698
1	87.9724	1	229.8633	1	371.754	1	513.645	1	655.536
2	90.8102	2	232.7012	2	374.592	2	516.483	2	658.374
3	93.6480	3	235.5390	3	377.430	3	519.321	3	661.212
4	96.4858	4	238.3768	4	380.268	4	522.159	4	664.050
5	99.3237	5	241.2146	5	383.106	5	524.996	5	666.887
6	102.1615	6	244.0524	6	385.943	6	527.834	6	669.725
7	104.9993	7	246.8902	7	388.781	7	530.672	7	672.563
8	107.8371	8	249.7281	8	391.619	8	533.510	8	675.401
9	110.6749	9	252.5659	9	394.457	9	536.348	9	678.239
40	113.5128	90	255.4037	140	397.295	190	539.186	240	681.077
1	116.3506	1	258.2415	1	400.132	1	542.023	1	683.914
2	119.1884	2	261.0793	2	402.970	2	544.861	2	686.752
3	122.0262	3	263.9172	3	405.808	3	547.699	3	689.590
4	124.8640	4	266.7550	4	408.646	4	550.537	4	692.428
5	127.7019	5	269.5928	5	411.484	5	553.375	5	695.266
6	130.5397	6	272.4306	6	414.322	6	556.213	6	698.103
7	133.3775	7	275.2684	7	417.159	7	559.050	7	700.941
8	136.2153	8	278.1063	8	419.997	8	561.888	8	703.779
9	139.0531	9	280.9441	9	422.835	9	564.726	9	706.617

IX. CAPACITY—HECTOLITRES TO BUSHELS

[Reduction factor: 1 hectolitre = 2.8378189 bushels]

Hecto-litres	Bushels	Hecto-litres	Bushels	Hecto-litres	Bushels	Hecto-litres	Bushels	Hecto-litres	Bushels
250	709.455	300	851.346	350	993.237	400	1,135.128	450	1,277.019
1	712.293	1	854.183	1	996.074	1	1,137.965	1	1,279.856
2	715.130	2	857.021	2	998.912	2	1,140.803	2	1,282.694
3	717.968	3	859.859	3	1,001.750	3	1,143.641	3	1,285.532
4	720.806	4	862.697	4	1,004.588	4	1,146.479	4	1,288.370
5	723.644	5	865.535	5	1,007.426	5	1,149.317	5	1,291.208
6	726.482	6	868.373	6	1,010.264	6	1,152.154	6	1,294.045
7	729.319	7	871.210	7	1,013.101	7	1,154.992	7	1,296.883
8	732.157	8	874.048	8	1,015.939	8	1,157.830	8	1,299.721
9	734.995	9	876.886	9	1,018.777	9	1,160.668	9	1,302.559
260	737.833	310	879.724	360	1,021.615	410	1,163.506	460	1,305.397
1	740.671	1	882.562	1	1,024.453	1	1,166.344	1	1,308.235
2	743.509	2	885.400	2	1,027.290	2	1,169.181	2	1,311.072
3	746.346	3	888.237	3	1,030.128	3	1,172.019	3	1,313.910
4	749.184	4	891.075	4	1,032.966	4	1,174.857	4	1,316.748
5	752.022	5	893.913	5	1,035.804	5	1,177.695	5	1,319.586
6	754.860	6	896.751	6	1,038.642	6	1,180.533	6	1,322.424
7	757.698	7	899.589	7	1,041.480	7	1,183.370	7	1,325.261
8	760.535	8	902.426	8	1,044.317	8	1,186.208	8	1,328.099
9	763.373	9	905.264	9	1,047.155	9	1,189.046	9	1,330.937
270	766.211	320	908.102	370	1,049.993	420	1,191.884	470	1,333.775
1	769.049	1	910.940	1	1,052.831	1	1,194.722	1	1,336.613
2	771.887	2	913.778	2	1,055.669	2	1,197.560	2	1,339.451
3	774.725	3	916.616	3	1,058.506	3	1,200.397	3	1,342.288
4	777.562	4	919.453	4	1,061.344	4	1,203.235	4	1,345.126
5	780.400	5	922.291	5	1,064.182	5	1,206.073	5	1,347.964
6	783.238	6	925.129	6	1,067.020	6	1,208.911	6	1,350.802
7	786.076	7	927.967	7	1,069.858	7	1,211.749	7	1,353.640
8	788.914	8	930.805	8	1,072.696	8	1,214.586	8	1,356.477
9	791.751	9	933.642	9	1,075.533	9	1,217.424	9	1,359.315
280	794.589	330	936.480	380	1,078.371	430	1,220.262	480	1,362.153
1	797.427	1	939.318	1	1,081.209	1	1,223.100	1	1,364.991
2	800.265	2	942.156	2	1,084.047	2	1,225.938	2	1,367.829
3	803.103	3	944.994	3	1,086.885	3	1,228.776	3	1,370.667
4	805.941	4	947.832	4	1,089.722	4	1,231.613	4	1,373.504
5	808.778	5	950.669	5	1,092.560	5	1,234.451	5	1,376.342
6	811.616	6	953.507	6	1,095.398	6	1,237.289	6	1,379.180
7	814.454	7	956.345	7	1,098.236	7	1,240.127	7	1,382.018
8	817.292	8	959.183	8	1,101.074	8	1,242.965	8	1,384.856
9	820.130	9	962.021	9	1,103.912	9	1,245.803	9	1,387.693
290	822.967	340	964.858	390	1,106.749	440	1,248.640	490	1,390.531
1	825.805	1	967.696	1	1,109.587	1	1,251.478	1	1,393.369
2	828.643	2	970.534	2	1,112.425	2	1,254.316	2	1,396.207
3	831.481	3	973.372	3	1,115.263	3	1,257.154	3	1,399.045
4	834.319	4	976.210	4	1,118.101	4	1,259.992	4	1,401.883
5	837.157	5	979.048	5	1,120.938	5	1,262.829	5	1,404.720
6	839.994	6	981.885	6	1,123.776	6	1,265.667	6	1,407.558
7	842.832	7	984.723	7	1,126.614	7	1,268.505	7	1,410.396
8	845.670	8	987.561	8	1,129.452	8	1,271.343	8	1,413.234
9	848.508	9	990.399	9	1,132.290	9	1,274.181	9	1,416.072

[Reduction factor: 1 hectolitre = 2.8378189 bushels]

Hecto-litres	Bushels	Hecto-litres	Bushels	Hecto-litres	Bushels	Hecto-litres	Bushels	Hecto-litres	Bushels
500	1,418.909	550	1,560.800	600	1,702.691	650	1,844.582	700	1,986.473
1	1,421.747	1	1,563.638	1	1,705.529	1	1,847.420	1	1,989.311
2	1,424.585	2	1,566.476	2	1,708.367	2	1,850.258	2	1,992.149
3	1,427.423	3	1,569.314	3	1,711.205	3	1,853.096	3	1,994.987
4	1,430.261	4	1,572.152	4	1,714.043	4	1,855.934	4	1,997.825
5	1,433.099	5	1,574.989	5	1,716.880	5	1,858.771	5	2,000.662
6	1,435.936	6	1,577.827	6	1,719.718	6	1,861.609	6	2,003.500
7	1,438.774	7	1,580.665	7	1,722.556	7	1,864.447	7	2,006.338
8	1,441.612	8	1,583.503	8	1,725.394	8	1,867.285	8	2,009.176
9	1,444.450	9	1,586.341	9	1,728.232	9	1,870.123	9	2,012.014
510	1,447.288	560	1,589.179	610	1,731.070	660	1,872.960	710	2,014.851
1	1,450.125	1	1,592.016	1	1,733.907	1	1,875.798	1	2,017.689
2	1,452.963	2	1,594.854	2	1,736.745	2	1,878.636	2	2,020.527
3	1,455.801	3	1,597.692	3	1,739.583	3	1,881.474	3	2,023.365
4	1,458.639	4	1,600.530	4	1,742.421	4	1,884.312	4	2,026.203
5	1,461.477	5	1,603.368	5	1,745.259	5	1,887.150	5	2,029.041
6	1,464.315	6	1,606.206	6	1,748.096	6	1,889.987	6	2,031.878
7	1,467.152	7	1,609.043	7	1,750.934	7	1,892.825	7	2,034.716
8	1,469.990	8	1,611.881	8	1,753.772	8	1,895.663	8	2,037.554
9	1,472.828	9	1,614.719	9	1,756.610	9	1,898.501	9	2,040.392
520	1,475.666	570	1,617.557	620	1,759.448	670	1,901.339	720	2,043.230
1	1,478.504	1	1,620.395	1	1,762.286	1	1,904.176	1	2,046.067
2	1,481.341	2	1,623.232	2	1,765.123	2	1,907.014	2	2,048.905
3	1,484.179	3	1,626.070	3	1,767.961	3	1,909.852	3	2,051.743
4	1,487.017	4	1,628.908	4	1,770.799	4	1,912.690	4	2,054.581
5	1,489.855	5	1,631.746	5	1,773.637	5	1,915.528	5	2,057.419
6	1,492.693	6	1,634.584	6	1,776.475	6	1,918.366	6	2,060.257
7	1,495.531	7	1,637.422	7	1,779.312	7	1,921.203	7	2,063.094
8	1,498.368	8	1,640.259	8	1,782.150	8	1,924.041	8	2,065.932
9	1,501.206	9	1,643.097	9	1,784.988	9	1,926.879	9	2,068.770
530	1,504.044	580	1,645.935	630	1,787.826	680	1,929.717	730	2,071.608
1	1,506.882	1	1,648.773	1	1,790.664	1	1,932.555	1	2,074.446
2	1,509.720	2	1,651.611	2	1,793.502	2	1,935.393	2	2,077.283
3	1,512.557	3	1,654.448	3	1,796.339	3	1,938.230	3	2,080.121
4	1,515.395	4	1,657.286	4	1,799.177	4	1,941.068	4	2,082.959
5	1,518.233	5	1,660.124	5	1,802.015	5	1,943.906	5	2,085.797
6	1,521.071	6	1,662.962	6	1,804.853	6	1,946.744	6	2,088.635
7	1,523.909	7	1,665.800	7	1,807.691	7	1,949.582	7	2,091.473
8	1,526.747	8	1,668.638	8	1,810.528	8	1,952.419	8	2,094.310
9	1,529.584	9	1,671.475	9	1,813.366	9	1,955.257	9	2,097.148
540	1,532.422	590	1,674.313	640	1,816.204	690	1,958.095	740	2,099.986
1	1,535.260	1	1,677.151	1	1,819.042	1	1,960.933	1	2,102.824
2	1,538.098	2	1,679.989	2	1,821.880	2	1,963.771	2	2,105.662
3	1,540.936	3	1,682.827	3	1,824.718	3	1,966.609	3	2,108.499
4	1,543.773	4	1,685.664	4	1,827.555	4	1,969.446	4	2,111.337
5	1,546.611	5	1,688.502	5	1,830.393	5	1,972.284	5	2,114.175
6	1,549.449	6	1,691.340	6	1,833.231	6	1,975.122	6	2,117.013
7	1,552.287	7	1,694.178	7	1,836.069	7	1,977.960	7	2,119.851
8	1,555.125	8	1,697.016	8	1,838.907	8	1,980.798	8	2,122.689
9	1,557.963	9	1,699.854	9	1,841.744	9	1,983.635	9	2,125.526

IX. CAPACITY—HECTOLITRES TO BUSHELS

[Reduction factor: 1 hectolitre = 2.8378189 bushels]

Hecto-litres	Bush-els	Hecto-litres	Bush-els	Hecto-litres	Bush-els	Hecto-litres	Bush-els	Hecto-litres	Bush-els
750	2,128.364	800	2,270.255	850	2,412.146	900	2,554.037	950	2,695.928
1	2,131.202	1	2,273.093	1	2,414.984	1	2,556.875	1	2,698.766
2	2,134.040	2	2,275.931	2	2,417.822	2	2,559.713	2	2,701.604
3	2,136.878	3	2,278.769	3	2,420.660	3	2,562.550	3	2,704.441
4	2,139.715	4	2,281.606	4	2,423.497	4	2,565.388	4	2,707.279
5	2,142.553	5	2,284.444	5	2,426.335	5	2,568.226	5	2,710.117
6	2,145.391	6	2,287.282	6	2,429.173	6	2,571.064	6	2,712.955
7	2,148.229	7	2,290.120	7	2,432.011	7	2,573.902	7	2,715.793
8	2,151.067	8	2,292.958	8	2,434.849	8	2,576.740	8	2,718.631
9	2,153.905	9	2,295.796	9	2,437.686	9	2,579.577	9	2,721.468
760	2,156.742	810	2,298.633	860	2,440.524	910	2,582.415	960	2,724.306
1	2,159.580	1	2,301.471	1	2,443.362	1	2,585.253	1	2,727.144
2	2,162.418	2	2,304.309	2	2,446.200	2	2,588.091	2	2,729.982
3	2,165.256	3	2,307.147	3	2,449.038	3	2,590.929	3	2,732.820
4	2,168.094	4	2,309.985	4	2,451.876	4	2,593.766	4	2,735.657
5	2,170.931	5	2,312.822	5	2,454.713	5	2,596.604	5	2,738.495
6	2,173.769	6	2,315.660	6	2,457.551	6	2,599.442	6	2,741.333
7	2,176.607	7	2,318.498	7	2,460.389	7	2,602.280	7	2,744.171
8	2,179.445	8	2,321.336	8	2,463.227	8	2,605.118	8	2,747.009
9	2,182.283	9	2,324.174	9	2,466.065	9	2,607.956	9	2,749.847
770	2,185.121	820	2,327.012	870	2,468.902	920	2,610.793	970	2,752.684
1	2,187.958	1	2,329.849	1	2,471.740	1	2,613.631	1	2,755.522
2	2,190.796	2	2,332.687	2	2,474.578	2	2,616.469	2	2,758.360
3	2,193.634	3	2,335.525	3	2,477.416	3	2,619.307	3	2,761.198
4	2,196.472	4	2,338.363	4	2,480.254	4	2,622.145	4	2,764.036
5	2,199.310	5	2,341.201	5	2,483.092	5	2,624.982	5	2,766.873
6	2,202.147	6	2,344.038	6	2,485.929	6	2,627.820	6	2,769.711
7	2,204.985	7	2,346.876	7	2,488.767	7	2,630.658	7	2,772.549
8	2,207.823	8	2,349.714	8	2,491.605	8	2,633.496	8	2,775.387
9	2,210.661	9	2,352.552	9	2,494.443	9	2,636.334	9	2,778.225
780	2,213.499	830	2,355.390	880	2,497.281	930	2,639.172	980	2,781.063
1	2,216.337	1	2,358.228	1	2,500.118	1	2,642.009	1	2,783.900
2	2,219.174	2	2,361.065	2	2,502.956	2	2,644.847	2	2,786.738
3	2,222.012	3	2,363.903	3	2,505.794	3	2,647.685	3	2,789.576
4	2,224.850	4	2,366.741	4	2,508.632	4	2,650.523	4	2,792.414
5	2,227.688	5	2,369.579	5	2,511.470	5	2,653.361	5	2,795.252
6	2,230.526	6	2,372.417	6	2,514.308	6	2,656.199	6	2,798.089
7	2,233.363	7	2,375.254	7	2,517.145	7	2,659.036	7	2,800.927
8	2,236.201	8	2,378.092	8	2,519.983	8	2,661.874	8	2,803.765
9	2,239.039	9	2,380.930	9	2,522.821	9	2,664.712	9	2,806.603
790	2,241.877	840	2,383.768	890	2,525.659	940	2,667.550	990	2,809.441
1	2,244.715	1	2,386.606	1	2,528.497	1	2,670.388	1	2,812.279
2	2,247.553	2	2,389.444	2	2,531.334	2	2,673.225	2	2,815.116
3	2,250.390	3	2,392.281	3	2,534.172	3	2,676.063	3	2,817.954
4	2,253.228	4	2,395.119	4	2,537.010	4	2,678.901	4	2,820.792
5	2,256.066	5	2,397.957	5	2,539.848	5	2,681.739	5	2,823.630
6	2,258.904	6	2,400.795	6	2,542.686	6	2,684.577	6	2,826.468
7	2,261.742	7	2,403.633	7	2,545.524	7	2,687.415	7	2,829.305
8	2,264.579	8	2,406.470	8	2,548.361	8	2,690.252	8	2,832.143
9	2,267.417	9	2,409.308	9	2,551.199	9	2,693.090	9	2,834.981

141

IX. CAPACITY—BUSHELS TO HECTOLITRES

[Reduction factor: 1 bushel = 0.35238330 hectolitre]

Bushels	Hectolitres	Bushels	Hectolitres	Bushels	Hectolitres	Bushels	Hectolitres	Bushels	Hectolitres
0		50	17.61917	100	35.2383	150	52.8575	200	70.4767
1	0.35238	1	17.97155	1	35.5907	1	53.2099	1	70.8290
2	0.70477	2	18.32393	2	35.9431	2	53.5623	2	71.1814
3	1.05715	3	18.67631	3	36.2955	3	53.9146	3	71.5338
4	1.40953	4	19.02870	4	36.6479	4	54.2670	4	71.8862
5	1.76192	5	19.38108	5	37.0002	5	54.6194	5	72.2386
6	2.11430	6	19.73346	6	37.3526	6	54.9718	6	72.5910
7	2.46668	7	20.08585	7	37.7050	7	55.3242	7	72.9433
8	2.81907	8	20.43823	8	38.0574	8	55.6766	8	73.2957
9	3.17145	9	20.79061	9	38.4098	9	56.0289	9	73.6481
10	3.52383	60	21.14300	110	38.7622	160	56.3813	210	74.0005
1	3.87622	1	21.49538	1	39.1145	1	56.7337	1	74.3529
2	4.22860	2	21.84776	2	39.4669	2	57.0861	2	74.7053
3	4.58098	3	22.20015	3	39.8193	3	57.4385	3	75.0576
4	4.93337	4	22.55253	4	40.1717	4	57.7909	4	75.4100
5	5.28575	5	22.90491	5	40.5241	5	58.1432	5	75.7624
6	5.63813	6	23.25730	6	40.8765	6	58.4956	6	76.1148
7	5.99052	7	23.60968	7	41.2288	7	58.8480	7	76.4672
8	6.34290	8	23.96206	8	41.5812	8	59.2004	8	76.8196
9	6.69528	9	24.31445	9	41.9336	9	59.5528	9	77.1719
20	7.04767	70	24.66683	120	42.2860	170	59.9052	220	77.5243
1	7.40005	1	25.01921	1	42.6384	1	60.2575	1	77.8767
2	7.75243	2	25.37160	2	42.9908	2	60.6099	2	78.2291
3	8.10482	3	25.72398	3	43.3431	3	60.9623	3	78.5815
4	8.45720	4	26.07636	4	43.6955	4	61.3147	4	78.9339
5	8.80958	5	26.42875	5	44.0479	5	61.6671	5	79.2862
6	9.16197	6	26.78113	6	44.4003	6	62.0195	6	79.6386
7	9.51435	7	27.13351	7	44.7527	7	62.3718	7	79.9910
8	9.86673	8	27.48590	8	45.1051	8	62.7242	8	80.3434
9	10.21912	9	27.83828	9	45.4574	9	63.0766	9	80.6958
30	10.57150	80	28.19066	130	45.8098	180	63.4290	230	81.0482
1	10.92388	1	28.54305	1	46.1622	1	63.7814	1	81.4005
2	11.27627	2	28.89543	2	46.5146	2	64.1338	2	81.7529
3	11.62865	3	29.24781	3	46.8670	3	64.4861	3	82.1053
4	11.98103	4	29.60020	4	47.2194	4	64.8385	4	82.4577
5	12.33342	5	29.95258	5	47.5717	5	65.1909	5	82.8101
6	12.68580	6	30.30496	6	47.9241	6	65.5433	6	83.1625
7	13.03818	7	30.65735	7	48.2765	7	65.8957	7	83.5148
8	13.39057	8	31.00973	8	48.6289	8	66.2481	8	83.8672
9	13.74295	9	31.36211	9	48.9813	9	66.6004	9	84.2196
40	14.09533	90	31.71450	140	49.3337	190	66.9528	240	84.5720
1	14.44772	1	32.06688	1	49.6860	1	67.3052	1	84.9244
2	14.80010	2	32.41926	2	50.0384	2	67.6576	2	85.2768
3	15.15248	3	32.77165	3	50.3908	3	68.0100	3	85.6291
4	15.50487	4	33.12403	4	50.7432	4	68.3624	4	85.9815
5	15.85725	5	33.47641	5	51.0956	5	68.7147	5	86.3339
6	16.20963	6	33.82880	6	51.4480	6	69.0671	6	86.6863
7	16.56202	7	34.18118	7	51.8003	7	69.4195	7	87.0387
8	16.91440	8	34.53356	8	52.1527	8	69.7719	8	87.3911
9	17.26678	9	34.88595	9	52.5051	9	70.1243	9	87.7434

IX. CAPACITY—BUSHELS TO HECTOLITRES

[Reduction factor: 1 bushel = 0.35238330 hectolitre]

Bush-els	Hecto-litres	Bush-els	Hecto-litres	Bush-els	Hecto-litres	Bush-els	Hecto-litres	Bush-els	Hecto-litres
250	88.0958	300	105.7150	350	123.3342	400	140.9533	450	158.5725
1	88.4482	1	106.0674	1	123.6865	1	141.3057	1	158.9249
2	88.8006	2	106.4198	2	124.0389	2	141.6581	2	159.2773
3	89.1530	3	106.7721	3	124.3913	3	142.0105	3	159.6296
4	89.5054	4	107.1245	4	124.7437	4	142.3629	4	159.9820
5	89.8577	5	107.4769	5	125.0961	5	142.7152	5	160.3344
6	90.2101	6	107.8293	6	125.4485	6	143.0676	6	160.6868
7	90.5625	7	108.1817	7	125.8008	7	143.4200	7	161.0392
8	90.9149	8	108.5341	8	126.1532	8	143.7724	8	161.3916
9	91.2673	9	108.8864	9	126.5056	9	144.1248	9	161.7439
260	91.6197	310	109.2388	360	126.8580	410	144.4772	460	162.0963
1	91.9720	1	109.5912	1	127.2104	1	144.8295	1	162.4487
2	92.3244	2	109.9436	2	127.5628	2	145.1819	2	162.8011
3	92.6768	3	110.2960	3	127.9151	3	145.5343	3	163.1535
4	93.0292	4	110.6484	4	128.2675	4	145.8867	4	163.5059
5	93.3816	5	111.0007	5	128.6199	5	146.2391	5	163.8582
6	93.7340	6	111.3531	6	128.9723	6	146.5915	6	164.2106
7	94.0863	7	111.7055	7	129.3247	7	146.9438	7	164.5630
8	94.4387	8	112.0579	8	129.6771	8	147.2962	8	164.9154
9	94.7911	9	112.4103	9	130.0294	9	147.6486	9	165.2678
270	95.1435	320	112.7627	370	130.3818	420	148.0010	470	165.6202
1	95.4959	1	113.1150	1	130.7342	1	148.3534	1	165.9725
2	95.8483	2	113.4674	2	131.0866	2	148.7058	2	166.3249
3	96.2006	3	113.8198	3	131.4390	3	149.0581	3	166.6773
4	96.5530	4	114.1722	4	131.7914	4	149.4105	4	167.0297
5	96.9054	5	114.5746	5	132.1437	5	149.7629	5	167.3821
6	97.2578	6	114.8770	6	132.4961	6	150.1153	6	167.7345
7	97.6102	7	115.2293	7	132.8485	7	150.4677	7	168.0868
8	97.9626	8	115.5817	8	133.2009	8	150.8201	8	168.4392
9	98.3149	9	115.9341	9	133.5533	9	151.1724	9	168.7916
280	98.6673	330	116.2865	380	133.9057	430	151.5248	480	169.1440
1	99.0197	1	116.6389	1	134.2580	1	151.8772	1	169.4964
2	99.3721	2	116.9913	2	134.6104	2	152.2296	2	169.8488
3	99.7245	3	117.3436	3	134.9628	3	152.5820	3	170.2011
4	100.0769	4	117.6960	4	135.3152	4	152.9344	4	170.5535
5	100.4292	5	118.0484	5	135.6676	5	153.2867	5	170.9059
6	100.7816	6	118.4008	6	136.0200	6	153.6391	6	171.2583
7	101.1340	7	118.7532	7	136.3723	7	153.9915	7	171.6107
8	101.4864	8	119.1056	8	136.7247	8	154.3439	8	171.9631
9	101.8388	9	119.4579	9	137.0771	9	154.6963	9	172.3154
290	102.1912	340	119.8103	390	137.4295	440	155.0487	490	172.6678
1	102.5435	1	120.1627	1	137.7819	1	155.4010	1	173.0202
2	102.8959	2	120.5151	2	138.1343	2	155.7534	2	173.3726
3	103.2483	3	120.8675	3	138.4866	3	156.1058	3	173.7250
4	103.6007	4	121.2199	4	138.8390	4	156.4582	4	174.0774
5	103.9531	5	121.5722	5	139.1914	5	156.8106	5	174.4297
6	104.3055	6	121.9246	6	139.5438	6	157.1630	6	174.7821
7	104.6578	7	122.2770	7	139.8962	7	157.5153	7	175.1345
8	105.0102	8	122.6294	8	140.2486	8	157.8677	8	175.4869
9	105.3626	9	122.9818	9	140.6009	9	158.2201	9	175.8393

IX. CAPACITY—BUSHELS TO HECTOLITRES

[Reduction factor: 1 bushel = 0.35238330 hectolitre]

Bush-els	Hecto-litres	Bush-els	Hecto-litres	Bush-els	Hecto-litres	Bush-els	Hecto-litres	Bush-els	Hecto-litres
500	176.1917	550	193.8108	600	211.4300	650	229.0491	700	246.6683
1	176.5440	1	194.1632	1	211.7824	1	229.4015	1	247.0207
2	176.8964	2	194.5156	2	212.1347	2	229.7539	2	247.3731
3	177.2488	3	194.8680	3	212.4871	3	230.1063	3	247.7255
4	177.6012	4	195.2203	4	212.8395	4	230.4587	4	248.0778
5	177.9536	5	195.5727	5	213.1919	5	230.8111	5	248.4302
6	178.3060	6	195.9251	6	213.5443	6	231.1634	6	248.7826
7	178.6583	7	196.2775	7	213.8967	7	231.5158	7	249.1350
8	179.0107	8	196.6299	8	214.2490	8	231.8682	8	249.4874
9	179.3631	9	196.9823	9	214.6014	9	232.2206	9	249.8398
510	179.7155	560	197.3346	610	214.9538	660	232.5730	710	250.1921
1	180.0679	1	197.6870	1	215.3062	1	232.9254	1	250.5445
2	180.4203	2	198.0394	2	215.6586	2	233.2777	2	250.8969
3	180.7726	3	198.3918	3	216.0110	3	233.6301	3	251.2493
4	181.1250	4	198.7442	4	216.3633	4	233.9825	4	251.6017
5	181.4774	5	199.0966	5	216.7157	5	234.3349	5	251.9541
6	181.8298	6	199.4489	6	217.0681	6	234.6873	6	252.3064
7	182.1822	7	199.8013	7	217.4205	7	235.0397	7	252.6588
8	182.5346	8	200.1537	8	217.7729	8	235.3920	8	253.0112
9	182.8869	9	200.5061	9	218.1253	9	235.7444	9	253.3636
520	183.2393	570	200.8585	620	218.4776	670	236.0968	720	253.7160
1	183.5917	1	201.2109	1	218.8300	1	236.4492	1	254.0684
2	183.9441	2	201.5632	2	219.1824	2	236.8016	2	254.4207
3	184.2965	3	201.9156	3	219.5348	3	237.1540	3	254.7731
4	184.6489	4	202.2680	4	219.8872	4	237.5063	4	255.1255
5	185.0012	5	202.6204	5	220.2396	5	237.8587	5	255.4779
6	185.3536	6	202.9728	6	220.5919	6	238.2111	6	255.8303
7	185.7060	7	203.3252	7	220.9443	7	238.5635	7	256.1827
8	186.0584	8	203.6775	8	221.2967	8	238.9159	8	256.5350
9	186.4108	9	204.0299	9	221.6491	9	239.2683	9	256.8874
530	186.7631	580	204.3823	630	222.0015	680	239.6206	730	257.2398
1	187.1155	1	204.7347	1	222.3539	1	239.9730	1	257.5922
2	187.4679	2	205.0871	2	222.7062	2	240.3254	2	257.9446
3	187.8203	3	205.4395	3	223.0586	3	240.6778	3	258.2970
4	188.1727	4	205.7918	4	223.4110	4	241.0302	4	258.6493
5	188.5251	5	206.1442	5	223.7634	5	241.3826	5	259 0017
6	188.8774	6	206.4966	6	224.1158	6	241.7349	6	259.3541
7	189.2298	7	206.8490	7	224.4682	7	242.0873	7	259.7065
8	189.5822	8	207.2014	8	224.8205	8	242.4397	8	260.0589
9	189.9346	9	207.5538	9	225.1729	9	242.7921	9	260.4113
540	190.2870	590	207.9061	640	225.5253	690	243.1445	740	260.7636
1	190.6394	1	208.2585	1	225.8777	1	243.4969	1	261.1160
2	190.9917	2	208.6109	2	226.2301	2	243.8492	2	261.4684
3	191.3441	3	208.9633	3	226.5825	3	244.2016	3	261.8208
4	191.6965	4	209.3157	4	226.9348	4	244.5540	4	262.1732
5	192.0489	5	209.6681	5	227.2872	5	244.9064	5	262.5256
6	192.4013	6	210.0204	6	227.6396	6	245.2588	6	262.8779
7	192.7537	7	210.3728	7	227.9920	7	245.6112	7	263.2303
8	193.1060	8	210.7252	8	228.3444	8	245.9635	8	263.5827
9	193.4584	9	211.0776	9	228.6968	9	246.3159	9	263.9351

[Reduction factor: 1 bushel = 0.35238330 hectolitre]

Bushels	Hectolitres	Bushels	Hectolitres	Bushels	Hectolitres	Bushels	Hectolitres	Bushels	Hectolitres
750	264.2875	800	281.9066	850	299.5258	900	317.1450	950	334.7641
1	264.6399	1	282.2590	1	299.8782	1	317.4974	1	335.1165
2	264.9922	2	282.6114	2	300.2306	2	317.8497	2	335.4689
3	265.3446	3	282.9638	3	300.5830	3	318.2021	3	335.8213
4	265.6970	4	283.3162	4	300.9353	4	318.5545	4	336.1737
5	266.0494	5	283.6686	5	301.2877	5	318.9069	5	336.5261
6	266.4018	6	284.0209	6	301.6401	6	319.2593	6	336.8784
7	266.7542	7	284.3733	7	301.9925	7	319.6117	7	337.2308
8	267.1065	8	284.7257	8	302.3449	8	319.9640	8	337.5832
9	267.4589	9	285.0781	9	302.6973	9	320.3164	9	337.9356
760	267.8113	810	285.4305	860	303.0496	910	320.6688	960	338.2880
1	268.1637	1	285.7829	1	303.4020	1	321.0212	1	338.6404
2	268.5161	2	286.1352	2	303.7544	2	321.3736	2	338.9927
3	268.8685	3	286.4876	3	304.1068	3	231.7260	3	339.3451
4	269.2208	4	286.8400	4	304.4592	4	322.0783	4	339.6975
5	269.5732	5	287.1924	5	304.8116	5	322.4307	5	340.0499
6	269.9256	6	287.5448	6	305.1639	6	322.7831	6	340.4023
7	270.2780	7	287.8972	7	305.5163	7	323.1355	7	340.7547
8	270.6304	8	288.2495	8	305.8687	8	323.4879	8	341.1070
9	270.9828	9	288.6019	9	306.2211	9	323.8403	9	341.4594
770	271.3351	820	288.9543	870	306.5735	920	324.1926	970	341.8118
1	271.6875	1	289.3067	1	306.9259	1	324.5450	1	342.1642
2	272.0399	2	289.6591	2	307.2782	2	324.8974	2	342.5166
3	272.3923	3	290.0115	3	307.6306	3	325.2498	3	342.8690
4	272.7447	4	290.3638	4	307.9830	4	325.6022	4	343.2213
5	273.0971	5	290.7162	5	308.3354	5	325.9546	5	343.5737
6	273.4494	6	291.0686	6	308.6878	6	326.3069	6	343.9261
7	273.8018	7	291.4210	7	309.0402	7	326.6593	7	344.2785
8	274.1542	8	291.7734	8	309.3925	8	327.0117	8	344.6309
9	274.5066	9	292.1258	9	309.7449	9	327.3641	9	344.9833
780	274.8590	830	292.4781	880	310.0973	930	327.7165	980	345.3356
1	275.2114	1	292.8305	1	310.4497	1	328.0689	1	345.6880
2	275.5637	2	293.1829	2	310.8021	2	328.4212	2	346.0404
3	275.9161	3	293.5353	3	311.1545	3	328.7736	3	346.3928
4	276.2685	4	293.8877	4	311.5068	4	329.1260	4	346.7452
5	276.6209	5	294.2401	5	311.8592	5	329.4784	5	347.0976
6	276.9733	6	294.5924	6	312.2116	6	329.8308	6	347.4499
7	277.3257	7	294.9448	7	312.5640	7	330.1832	7	347.8023
8	277.6780	8	295.2972	8	312.9164	8	330.5355	8	348.1547
9	278.0304	9	295.6496	9	313.2688	9	330.8879	9	348.5071
790	278.3828	840	296.0020	890	313.6211	940	331.2403	990	348.8595
1	278.7352	1	296.3544	1	313.9735	1	331.5927	1	349.2119
2	279.0876	2	296.7067	2	314.3259	2	331.9451	2	349.5642
3	279.4400	3	297.0591	3	314.6783	3	332.2975	3	349.9166
4	279.7923	4	297.4115	4	315.0307	4	332.6498	4	350.2690
5	280.1447	5	297.7639	5	315.3831	5	333.0022	5	350.6214
6	280.4971	6	298.1163	6	315.7354	6	333.3546	6	350.9738
7	280.8495	7	298.4687	7	316.0878	7	333.7070	7	351.3262
8	281.2019	8	298.8210	8	316.4402	8	334.0594	8	351.6785
9	281.5543	9	299.1734	9	316.7926	9	334.4118	9	352.0309

IX. MASS—KILOGRAMS TO AVOIRDUPOIS POUNDS

[Reduction factor: 1 kilogram = 2.204622341 avoirdupois pounds]

Kilos	Pounds	Kilos	Pounds	Kilos	Pounds	Kilos	Pounds	Kilos	Pounds
0		50	110.2311	100	220.4622	150	330.6934	200	440.9245
1	2.2046	1	112.4357	1	222.6669	1	332.8980	1	443.1291
2	4.4092	2	114.6404	2	224.8715	2	335.1026	2	445.3337
3	6.6139	3	116.8450	3	227.0761	3	337.3072	3	447.5383
4	8.8185	4	119.0496	4	229.2807	4	339.5118	4	449.7430
5	11.0231	5	121.2542	5	231.4853	5	341.7165	5	451.9476
6	13.2277	6	123.4589	6	233.6900	6	343.9211	6	454.1522
7	15.4324	7	125.6635	7	235.8946	7	346.1257	7	456.3568
8	17.6370	8	127.8681	8	238.0992	8	348.3303	8	458.5614
9	19.8416	9	130.0727	9	240.3038	9	350.5350	9	460.7661
10	22.0462	60	132.2773	110	242.5085	160	352.7396	210	462.9707
1	24.2508	1	134.4820	1	244.7131	1	354.9442	1	465.1753
2	26.4555	2	136.6866	2	246.9177	2	357.1488	2	467.3799
3	28.6601	3	138.8912	3	249.1223	3	359.3534	3	469.5846
4	30.8647	4	141.0958	4	251.3269	4	361.5581	4	471.7892
5	33.0693	5	143.3005	5	253.5316	5	363.7627	5	473.9938
6	35.2740	6	145.5051	6	255.7362	6	365.9673	6	476.1984
7	37.4786	7	147.7097	7	257.9408	7	368.1719	7	478.4030
8	39.6832	8	149.9143	8	260.1454	8	370.3766	8	480.6077
9	41.8878	9	152.1189	9	262.3501	9	372.5812	9	482.8123
20	44.0924	70	154.3236	120	264.5547	170	374.7858	220	485.0169
1	46.2971	1	156.5282	1	266.7593	1	376.9904	1	487.2215
2	48.5017	2	158.7328	2	268.9639	2	379.1950	2	489.4262
3	50.7063	3	160.9374	3	271.1685	3	381.3997	3	491.6308
4	52.9109	4	163.1421	4	273.3732	4	383.6043	4	493.8354
5	55.1156	5	165.3467	5	275.5778	5	385.8089	5	496.0400
6	57.3202	6	167.5513	6	277.7824	6	388.0135	6	498.2446
7	59.5248	7	169.7559	7	279.9870	7	390.2182	7	500.4493
8	61.7294	8	171.9605	8	282.1917	8	392.4228	8	502.6539
9	63.9340	9	174.1652	9	284.3963	9	394.6274	9	504.8585
30	66.1387	80	176.3698	130	286.6009	180	396.8320	230	507.0631
1	68.3433	1	178.5744	1	288.8055	1	399.0366	1	509.2678
2	70.5479	2	180.7790	2	291.0101	2	401.2413	2	511.4724
3	72.7525	3	182.9837	3	293.2148	3	403.4459	3	513.6770
4	74.9572	4	185.1883	4	295.4194	4	405.6505	4	515.8816
5	77.1618	5	187.3929	5	297.6240	5	407.8551	5	518.0863
6	79.3664	6	189.5975	6	299.8286	6	410.0598	6	520.2909
7	81.5710	7	191.8021	7	302.0333	7	412.2644	7	522.4955
8	83.7756	8	194.0068	8	304.2379	8	414.4690	8	524.7001
9	85.9803	9	196.2114	9	306.4425	9	416.6736	9	526.9047
40	88.1849	90	198.4160	140	308.6471	190	418.8782	240	529.1094
1	90.3895	1	200.6206	1	310.8518	1	421.0829	1	531.3140
2	92.5941	2	202.8253	2	313.0564	2	423.2875	2	533.5186
3	94.7988	3	205.0299	3	315.2610	3	425.4921	3	535.7232
4	97.0034	4	207.2345	4	317.4656	4	427.6967	4	537.9279
5	99.2080	5	209.4391	5	319.6702	5	429.9014	5	540.1325
6	101.4126	6	211.6437	6	321.8749	6	432.1060	6	542.3371
7	103.6173	7	213.8484	7	324.0795	7	434.3106	7	544.5417
8	105.8219	8	216.0530	8	326.2841	8	436.5152	8	546.7463
9	108.0265	9	218.2576	9	328.4887	9	438.7198	9	548.9510

IX. MASS—KILOGRAMS TO AVOIRDUPOIS POUNDS

[Reduction factor: 1 kilogram = 2.204622341 avoirdupois pounds]

Kilos	Pounds	Kilos	Pounds	Kilos	Pounds	Kilos	Pounds	Kilos	Pounds
250	551.1556	300	661.3867	350	771.6178	400	881.8489	450	992.0801
1	553.3602	1	663.5913	1	773.8224	1	884.0536	1	994.2847
2	555.5648	2	665.7959	2	776.0271	2	886.2582	2	996.4893
3	557.7695	3	668.0006	3	778.2317	3	888.4628	3	998.6939
4	559.9741	4	670.2052	4	780.4363	4	890.6674	4	1,000.8985
5	562.1787	5	672.4098	5	782.6409	5	892.8720	5	1,003.1032
6	564.3833	6	674.6144	6	784.8456	6	895.0767	6	1,005.3078
7	566.5879	7	676.8191	7	787.0502	7	897.2813	7	1,007.5124
8	568.7926	8	679.0237	8	789.2548	8	899.4859	8	1,009.7170
9	570.9972	9	681.2283	9	791.4594	9	901.6905	9	1,011.9217
260	573.2018	310	683.4329	360	793.6640	410	903.8952	460	1,014.1263
1	575.4064	1	685.6375	1	795.8687	1	906.0998	1	1,016.3309
2	577.6111	2	687.8422	2	798.0733	2	908.3044	2	1,018.5355
3	579.8157	3	690.0468	3	800.2779	3	910.5090	3	1,020.7401
4	582.0203	4	692.2514	4	802.4825	4	912.7136	4	1,022.9448
5	584.2249	5	694.4560	5	804.6872	5	914.9183	5	1,025.1494
6	586.4295	6	696.6607	6	806.8918	6	917.1229	6	1,027.3540
7	588.6342	7	698.8653	7	809.0964	7	919.3275	7	1,029.5586
8	590.8388	8	701.0699	8	811.3010	8	921.5321	8	1,031.7633
9	593.0434	9	703.2745	9	813.5056	9	923.7368	9	1,033.9679
270	595.2480	320	705.4791	370	815.7103	420	925.9414	470	1,036.1725
1	597.4527	1	707.6838	1	817.9149	1	928.1460	1	1,038.3771
2	599.6573	2	709.8884	2	820.1195	2	930.3506	2	1,040.5817
3	601.8619	3	712.0930	3	822.3241	3	932.5553	3	1,042.7864
4	604.0665	4	714.2976	4	824.5288	4	934.7599	4	1,044.9910
5	606.2711	5	716.5023	5	826.7334	5	936.9645	5	1,047.1956
6	608.4758	6	718.7069	6	828.9380	6	939.1691	6	1,049.4002
7	610.6804	7	720.9115	7	831.1426	7	941.3737	7	1,051.6049
8	612.8850	8	723.1161	8	833.3472	8	943.5784	8	1,053.8095
9	615.0896	9	725.3208	9	835.5519	9	945.7830	9	1,056.0141
280	617.2943	330	727.5254	380	837.7565	430	947.9876	480	1,058.2187
1	619.4989	1	729.7300	1	839.9611	1	950.1922	1	1,060.4233
2	621.7035	2	731.9346	2	842.1657	2	952.3969	2	1,062.6280
3	623.9081	3	734.1392	3	844.3704	3	954.6015	3	1,064.8326
4	626.1127	4	736.3439	4	846.5750	4	956.8061	4	1,067.0372
5	628.3174	5	738.5485	5	848.7796	5	959.0107	5	1,069.2418
6	630.5220	6	740.7531	6	850.9842	6	961.2153	6	1,071.4465
7	632.7266	7	742.9577	7	853.1888	7	963.4200	7	1,073.6511
8	634.9312	8	745.1624	8	855.3935	8	965.6246	8	1,075.8557
9	637.1359	9	747.3670	9	857.5981	9	967.8292	9	1,078.0603
290	639.3405	340	749.5716	390	859.8027	440	970.0338	490	1,080.2649
1	641.5451	1	751.7762	1	862.0073	1	972.2385	1	1,082.4696
2	643.7497	2	753.9808	2	864.2120	2	974.4431	2	1,084.6742
3	645.9543	3	756.1855	3	866.4166	3	976.6477	3	1,086.8788
4	648.1590	4	758.3901	4	868.6212	4	978.8523	4	1,089.0834
5	650.3636	5	760.5947	5	870.8258	5	981.0569	5	1,091.2881
6	652.5682	6	762.7993	6	873.0304	6	983.2616	6	1,093.4927
7	654.7728	7	765.0040	7	875.2351	7	985.4662	7	1,095.6973
8	656.9775	8	767.2086	8	877.4397	8	987.6708	8	1,097.9019
9	659.1821	9	769.4132	9	879.6443	9	989.8754	9	1,100.1065

IX. MASS—KILOGRAMS TO AVOIRDUPOIS POUNDS

[Reduction factor: 1 kilogram = 2.204622341 avoirdupois pounds]

Kilos	Pounds	Kilos	Pounds	Kilos	Pounds	Kilos	Pounds	Kilos	Pounds
500	1,102.3112	550	1,212.5423	600	1,322.7734	650	1,433.0045	700	1,543.2356
1	1,104.5158	1	1,214.7469	1	1,324.9780	1	1,435.2091	1	1,545.4403
2	1,106.7204	2	1,216.9515	2	1,327.1826	2	1,437.4138	2	1,547.6449
3	1,108.9250	3	1,219.1562	3	1,329.3873	3	1,439.6184	3	1,549.8495
4	1,111.1297	4	1,221.3608	4	1,331.5919	4	1,441.8230	4	1,552.0541
5	1,113.3343	5	1,223.5654	5	1,333.7965	5	1,444.0276	5	1,554.2588
6	1,115.5389	6	1,225.7700	6	1,336.0011	6	1,446.2323	6	1,556.4634
7	1,117.7435	7	1,227.9746	7	1,338.2058	7	1,448.4369	7	1,558.6680
8	1,119.9481	8	1,230.1793	8	1,340.4104	8	1,450.6415	8	1,560.8726
9	1,122.1528	9	1,232.3839	9	1,342.6150	9	1,452.8461	9	1,563.0772
510	1,124.3574	560	1,234.5885	610	1,344.8196	660	1,455.0507	710	1,565.2819
1	1,126.5620	1	1,236.7931	1	1,347.0243	1	1,457.2554	1	1,567.4865
2	1,128.7666	2	1,238.9978	2	1,349.2289	2	1,459.4600	2	1,569.6911
3	1,130.9713	3	1,241.2024	3	1,351.4335	3	1,461.6646	3	1,571.8957
4	1,133.1759	4	1,243.4070	4	1,353.6381	4	1,463.8692	4	1,574.1004
5	1,135.3805	5	1,245.6116	5	1,355.8427	5	1,466.0739	5	1,576.3050
6	1,137.5851	6	1,247.8162	6	1,358.0474	6	1,468.2785	6	1,578.5096
7	1,139.7898	7	1,250.0209	7	1,360.2520	7	1,470.4831	7	1,580.7142
8	1,141.9944	8	1,252.2255	8	1,362.4566	8	1,472.6877	8	1,582.9188
9	1,144.1990	9	1,254.4301	9	1,364.6612	9	1,474.8923	9	1,585.1235
520	1,146.4036	570	1,256.6347	620	1,366.8659	670	1,477.0970	720	1,587.3281
1	1,148.6082	1	1,258.8394	1	1,369.0705	1	1,479.3016	1	1,589.5327
2	1,150.8129	2	1,261.0440	2	1,371.2751	2	1,481.5062	2	1,591.7373
3	1,153.0175	3	1,263.2486	3	1,373.4797	3	1,483.7108	3	1,593.9420
4	1,155.2221	4	1,265.4532	4	1,375.6843	4	1,485.9155	4	1,596.1466
5	1,157.4267	5	1,267.6578	5	1,377.8890	5	1,488.1201	5	1,598.3512
6	1,159.6314	6	1,269.8625	6	1,380.0936	6	1,490.3247	6	1,600.5558
7	1,161.8360	7	1,272.0671	7	1,382.2982	7	1,492.5293	7	1,602.7604
8	1,164.0406	8	1,274.2717	8	1,384.5028	8	1,494.7339	8	1,604.9651
9	1,166.2452	9	1,276.4763	9	1,386.7075	9	1,496.9386	9	1,607.1697
530	1,168.4498	580	1,278.6810	630	1,388.9121	680	1,499.1432	730	1,609.3743
1	1,170.6545	1	1,280.8856	1	1,391.1167	1	1,501.3478	1	1,611.5789
2	1,172.8591	2	1,283.0902	2	1,393.3213	2	1,503.5524	2	1,613.7836
3	1,175.0637	3	1,285.2948	3	1,395.5259	3	1,505.7571	3	1,615.9882
4	1,177.2683	4	1,287.4994	4	1,397.7306	4	1,507.9617	4	1,618.1928
5	1,179.4730	5	1,289.7041	5	1,399.9352	5	1,510.1663	5	1,620.3974
6	1,181.6776	6	1,291.9087	6	1,402.1398	6	1,512.3709	6	1,622.6020
7	1,183.8822	7	1,294.1133	7	1,404.3444	7	1,514.5755	7	1,624.8067
8	1,186.0868	8	1,296.3179	8	1,406.5491	8	1,516.7802	8	1,627.0113
9	1,188.2914	9	1,298.5226	9	1,408.7537	9	1,518.9848	9	1,629.2159
540	1,190.4961	590	1,300.7272	640	1,410.9583	690	1,521.1894	740	1,631.4205
1	1,192.7007	1	1,302.9318	1	1,413.1629	1	1,523.3940	1	1,633.6252
2	1,194.9053	2	1,305.1364	2	1,415.3675	2	1,525.5987	2	1,635.8298
3	1,197.1099	3	1,307.3410	3	1,417.5722	3	1,527.8033	3	1,638.0344
4	1,199.3146	4	1,309.5457	4	1,419.7768	4	1,530.0079	4	1,640.2390
5	1,201.5192	5	1,311.7503	5	1,421.9814	5	1,532.2125	5	1,642.4436
6	1,203.7238	6	1,313.9549	6	1,424.1860	6	1,534.4171	6	1,644.6483
7	1,205.9284	7	1,316.1595	7	1,426.3907	7	1,536.6218	7	1,646.8529
8	1,208.1330	8	1,318.3642	8	1,428.5953	8	1,538.8264	8	1,649.0575
9	1,210.3377	9	1,320.5688	9	1,430.7999	9	1,541.0310	9	1,651.2621

IX. MASS—KILOGRAMS TO AVOIRDUPOIS POUNDS

[Reduction factor: 1 kilogram = 2.204622341 avoirdupois pounds]

Kilos	Pounds	Kilos	Pounds	Kilos	Pounds	Kilos	Pounds	Kilos	Pounds
750	1,653.4668	800	1,763.6979	850	1,873.9290	900	1,984.1601	950	2,094.3912
1	1,655.6714	1	1,765.9025	1	1,876.1336	1	1,986.3647	1	2,096.5958
2	1,657.8760	2	1,768.1071	2	1,878.3382	2	1,988.5694	2	2,098.8005
3	1,660.0806	3	1,770.3117	3	1,880.5429	3	1,990.7740	3	2,101.0051
4	1,662.2852	4	1,772.5164	4	1,882.7475	4	1,992.9786	4	2,103.2097
5	1,664.4899	5	1,774.7210	5	1,884.9521	5	1,995.1832	5	2,105.4143
6	1,666.6945	6	1,776.9256	6	1,887.1567	6	1,997.3878	6	2,107.6190
7	1,668.8991	7	1,779.1302	7	1,889.3613	7	1,999.5925	7	2,109.8236
8	1,671.1037	8	1,781.3349	8	1,891.5660	8	2,001.7971	8	2,112.0282
9	1,673.3084	9	1,783.5395	9	1,893.7706	9	2,004.0017	9	2,114.2328
760	1,675.5130	810	1,785.7441	860	1,895.9752	910	2,006.2063	960	2,116.4374
1	1,677.7176	1	1,787.9487	1	1,898.1798	1	2,008.4110	1	2,118.6421
2	1,679.9222	2	1,790.1533	2	1,900.3845	2	2,010.6156	2	2,120.8467
3	1,682.1268	3	1,792.3580	3	1,902.5891	3	2,012.8202	3	2,123.0513
4	1,684.3315	4	1,794.5626	4	1,904.7937	4	2,015.0248	4	2,125.2559
5	1,686.5361	5	1,796.7672	5	1,906.9983	5	2,017.2294	5	2,127.4606
6	1,688.7407	6	1,798.9718	6	1,909.2029	6	2,019.4341	6	2,129.6652
7	1,690.9453	7	1,801.1765	7	1,911.4076	7	2,021.6387	7	2,131.8698
8	1,693.1500	8	1,803.3811	8	1,913.6122	8	2,023.8433	8	2,134.0744
9	1,695.3546	9	1,805.5857	9	1,915.8168	9	2,026.0479	9	2,136.2790
770	1,697.5592	820	1,807.7903	870	1,918.0214	920	2,028.2526	970	2,138.4837
1	1,699.7638	1	1,809.9949	1	1,920.2261	1	2,030.4572	1	2,140.6883
2	1,701.9684	2	1,812.1996	2	1,922.4307	2	2,032.6618	2	2,142.8929
3	1,704.1731	3	1,814.4042	3	1,924.6353	3	2,034.8664	3	2,145.0975
4	1,706.3777	4	1,816.6088	4	1,926.8399	4	2,037.0710	4	2,147.3022
5	1,708.5823	5	1,818.8134	5	1,929.0445	5	2,039.2757	5	2,149.5068
6	1,710.7869	6	1,821.0181	6	1,931.2492	6	2,041.4803	6	2,151.7114
7	1,712.9916	7	1,823.2227	7	1,933.4538	7	2,043.6849	7	2,153.9160
8	1,715.1962	8	1,825.4273	8	1,935.6584	8	2,045.8895	8	2,156.1206
9	1,717.4008	9	1,827.6319	9	1,937.8630	9	2,048.0942	9	2,158.3258
780	1,719.6054	830	1,829.8365	880	1,940.0677	930	2,050.2988	980	2,160.5299
1	1,721.8100	1	1,832.0412	1	1,942.2723	1	2,052.5034	1	2,162.7345
2	1,724.0147	2	1,834.2458	2	1,944.4769	2	2,054.7080	2	2,164.9391
3	1,726.2193	3	1,836.4504	3	1,946.6815	3	2,056.9126	3	2,167.1438
4	1,728.4239	4	1,838.6550	4	1,948.8861	4	2,059.1173	4	2,169.3484
5	1,730.6285	5	1,840.8597	5	1,951.0908	5	2,061.3219	5	2,171.5530
6	1,732.8332	6	1,843.0643	6	1,953.2954	6	2,063.5265	6	2,173.7576
7	1,735.0378	7	1,845.2689	7	1,955.5000	7	2,065.7311	7	2,175.9623
8	1,737.2424	8	1,847.4735	8	1,957.7046	8	2,067.9358	8	2,178.1669
9	1,739.4470	9	1,849.6781	9	1,959.9093	9	2,070.1404	9	2,180.3715
790	1,741.6516	840	1,851.8828	890	1,962.1139	940	2,072.3450	990	2,182.5761
1	1,743.8563	1	1,854.0874	1	1,964.3185	1	2,074.5496	1	2,184.7807
2	1,746.0609	2	1,856.2920	2	1,966.5231	2	2,076.7542	2	2,186.9854
3	1,748.2655	3	1,858.4966	3	1,968.7278	3	2,078.9589	3	2,189.1900
4	1,750.4701	4	1,860.7013	4	1,970.9324	4	2,081.1635	4	2,191.3946
5	1,752.6748	5	1,862.9059	5	1,973.1370	5	2,083.3681	5	2,193.5992
6	1,754.8794	6	1,865.1105	6	1,975.3416	6	2,085.5727	6	2,195.8039
7	1,757.0840	7	1,867.3151	7	1,977.5462	7	2,087.7774	7	2,198.0085
8	1,759.2886	8	1,869.5197	8	1,979.7509	8	2,089.9820	8	2,200.2131
9	1,761.4933	9	1,871.7244	9	1,981.9555	9	2,092.1866	9	2,202.4177

IX. MASS—AVOIRDUPOIS POUNDS TO KILOGRAMS

[Reduction factor: 1 avoirdupois pound = 0.4535924277 kilogram]

Pounds	Kilos	Pounds	Kilos	Pounds	Kilos	Pounds	Kilos	Pounds	Kilos
0		50	22.67962	100	45.35924	150	68.03886	200	90.71849
1	0.45359	1	23.13321	1	45.81284	1	68.49246	1	91.17208
2	.90718	2	23.58681	2	46.26643	2	68.94605	2	91.62567
3	1.36078	3	24.04040	3	46.72002	3	69.39964	3	92.07926
4	1.81437	4	24.49399	4	47.17361	4	69.85323	4	92.53286
5	2.26796	5	24.94758	5	47.62720	5	70.30683	5	92.98645
6	2.72155	6	25.40118	6	48.08080	6	70.76042	6	93.44004
7	3.17515	7	25.85477	7	48.53439	7	71.21401	7	93.89363
8	3.62874	8	26.30836	8	48.98798	8	71.66760	8	94.34722
9	4.08233	9	26.76195	9	49.44157	9	72.12120	9	94.80082
10	4.53592	60	27.21555	110	49.89517	160	72.57479	210	95.25441
1	4.98952	1	27.66914	1	50.34876	1	73.02838	1	95.70800
2	5.44311	2	28.12273	2	50.80235	2	73.48197	2	96.16159
3	5.89670	3	28.57632	3	51.25594	3	73.93557	3	96.61519
4	6.35029	4	29.02992	4	51.70954	4	74.38916	4	97.06878
5	6.80389	5	29.48351	5	52.16313	5	74.84275	5	97.52237
6	7.25748	6	29.93710	6	52.61672	6	75.29634	6	97.97596
7	7.71107	7	30.39069	7	53.07031	7	75.74994	7	98.42956
8	8.16466	8	30.84429	8	53.52391	8	76.20353	8	98.88315
9	8.61826	9	31.29788	9	53.97750	9	76.65712	9	99.33674
20	9.07185	70	31.75147	120	54.43109	170	77.11071	220	99.79033
1	9.52544	1	32.20506	1	54.88468	1	77.56431	1	100.24393
2	9.97903	2	32.65865	2	55.33828	2	78.01790	2	100.69752
3	10.43263	3	33.11225	3	55.79187	3	78.47149	3	101.15111
4	10.88622	4	33.56584	4	56.24546	4	78.92509	4	101.60470
5	11.33981	5	34.01943	5	56.69905	5	79.37867	5	102.05830
6	11.79340	6	34.47302	6	57.15265	6	79.83227	6	102.51189
7	12.24700	7	34.92662	7	57.60624	7	80.28586	7	102.96548
8	12.70059	8	35.38021	8	58.05983	8	80.73945	8	103.41907
9	13.15418	9	35.83380	9	58.51342	9	81.19304	9	103.87267
30	13.60777	80	36.28739	130	58.96702	180	81.64664	230	104.32626
1	14.06137	1	36.74099	1	59.42061	1	82.10023	1	104.77985
2	14.51496	2	37.19458	2	59.87420	2	82.55382	2	105.23344
3	14.96855	3	37.64817	3	60.32779	3	83.00741	3	105.68704
4	15.42214	4	38.10176	4	60.78139	4	83.46101	4	106.14063
5	15.87573	5	38.55536	5	61.23498	5	83.91460	5	106.59422
6	16.32933	6	39.00895	6	61.68857	6	84.36819	6	107.04781
7	16.78292	7	39.46254	7	62.14216	7	84.82178	7	107.50141
8	17.23651	8	39.91613	8	62.59576	8	85.27538	8	107.95500
9	17.69010	9	40.36973	9	63.04935	9	85.72897	9	108.40859
40	18.14370	90	40.82332	140	63.50294	190	86.18256	240	108.86218
1	18.59729	1	41.27691	1	63.95653	1	86.63615	1	109.31578
2	19.05088	2	41.73050	2	64.41012	2	87.08975	2	109.76937
3	19.50447	3	42.18410	3	64.86372	3	87.54334	3	110.22296
4	19.95807	4	42.63769	4	65.31731	4	87.99693	4	110.67655
5	20.41166	5	43.09128	5	65.77090	5	88.45052	5	111.13014
6	20.86525	6	43.54487	6	66.22449	6	88.90412	6	111.58374
7	21.31884	7	43.99847	7	66.67809	7	89.35771	7	112.03733
8	21.77244	8	44.45206	8	67.13168	8	89.81130	8	112.49092
9	22.22603	9	44.90565	9	67.58527	9	90.26489	9	112.94451

IX. MASS—AVOIRDUPOIS POUNDS TO KILOGRAMS

[Reduction factor: 1 avoirdupois pound = 0.4535924277 kilogram]

Pounds	Kilos	Pounds	Kilos	Pounds	Kilos	Pounds	Kilos	Pounds	Kilos
250	113.39811	300	136.07773	350	158.75735	400	181.43697	450	204.11659
1	113.85170	1	136.53132	1	159.21094	1	181.89056	1	204.57018
2	114.30529	2	136.98491	2	159.66453	2	182.34416	2	205.02378
3	114.75888	3	137.43851	3	160.11813	3	182.79775	3	205.47737
4	115.21248	4	137.89210	4	160.57172	4	183.25134	4	205.93096
5	115.66607	5	138.34569	5	161.02531	5	183.70493	5	206.38455
6	116.11966	6	138.79928	6	161.47890	6	184.15853	6	206.83815
7	116.57325	7	139.25288	7	161.93250	7	184.61212	7	207.29174
8	117.02685	8	139.70647	8	162.38609	8	185.06571	8	207.74533
9	117.48044	9	140.16006	9	162.83968	9	185.51930	9	208.19892
260	117.93403	310	140.61365	360	163.29327	410	185.97290	460	208.65252
1	118.38762	1	141.06725	1	163.74687	1	186.42649	1	209.10611
2	118.84122	2	141.52084	2	164.20046	2	186.88008	2	209.55970
3	119.29481	3	141.97443	3	164.65405	3	187.33367	3	210.01329
4	119.74840	4	142.42802	4	165.10764	4	187.78727	4	210.46689
5	120.20199	5	142.88161	5	165.56124	5	188.24086	5	210.92048
6	120.65559	6	143.33521	6	166.01483	6	188.69445	6	211.37407
7	121.10918	7	143.78880	7	166.46842	7	189.14804	7	211.82766
8	121.56277	8	144.24239	8	166.92201	8	189.60163	8	212.28126
9	122.01636	9	144.69598	9	167.37561	9	190.05523	9	212.73485
270	122.46996	320	145.14958	370	167.82920	420	190.50882	470	213.18844
1	122.92355	1	145.60317	1	168.28279	1	190.96241	1	213.64203
2	123.37714	2	146.05676	2	168.73638	2	191.41600	2	214.09563
3	123.83073	3	146.51035	3	169.18998	3	191.86960	3	214.54922
4	124.28433	4	146.96395	4	169.64357	4	192.32319	4	215.00281
5	124.73792	5	147.41754	5	170.09716	5	192.77678	5	215.45640
6	125.19151	6	147.87113	6	170.55075	6	193.23037	6	215.91000
7	125.64510	7	148.32472	7	171.00435	7	193.68397	7	216.36359
8	126.09869	8	148.77832	8	171.45794	8	194.13756	8	216.81718
9	126.55229	9	149.23191	9	171.91153	9	194.59115	9	217.27077
280	127.00588	330	149.68550	380	172.36512	430	195.04474	480	217.72437
1	127.45947	1	150.13909	1	172.81871	1	195.49834	1	218.17796
2	127.91306	2	150.59269	2	173.27231	2	195.95193	2	218.63155
3	128.36666	3	151.04628	3	173.72590	3	196.40552	3	219.08514
4	128.82025	4	151.49987	4	174.17949	4	196.85911	4	219.53874
5	129.27384	5	151.95346	5	174.63308	5	197.31271	5	219.99233
6	129.72743	6	152.40706	6	175.08668	6	197.76630	6	220.44592
7	130.18103	7	152.86065	7	175.54027	7	198.21989	7	220.89951
8	130.63462	8	153.31424	8	175.99386	8	198.67348	8	221.35310
9	131.08821	9	153.76783	9	176.44745	9	199.12708	9	221.80670
290	131.54180	340	154.22143	390	176.90105	440	199.58067	490	222.26029
1	131.99540	1	154.67502	1	177.35464	1	200.03426	1	222.71388
2	132.44899	2	155.12861	2	177.80823	2	200.48785	2	223.16747
3	132.90258	3	155.58220	3	178.26182	3	200.94145	3	223.62107
4	133.35617	4	156.03580	4	178.71542	4	201.39504	4	224.07466
5	133.80977	5	156.48939	5	179.16901	5	201.84863	5	224.52825
6	134.26336	6	156.94298	6	179.62260	6	202.30222	6	224.98184
7	134.71695	7	157.39657	7	180.07619	7	202.75582	7	225.43544
8	135.17054	8	157.85016	8	180.52979	8	203.20941	8	225.88903
9	135.62414	9	158.30376	9	180.98338	9	203.66300	9	226.34262

IX. MASS—AVOIRDUPOIS POUNDS TO KILOGRAMS

[Reduction factor: 1 avoirdupois pound = 0.4535924277 kilogram]

Pounds	Kilos	Pounds	Kilos	Pounds	Kilos	Pounds	Kilos	Pounds	Kilos
500	226.79621	550	249.47584	600	272.15546	650	294.83508	700	317.51470
1	227.24981	1	249.92943	1	272.60905	1	295.28867	1	317.96829
2	227.70340	2	250.38302	2	273.06264	2	295.74226	2	318.42188
3	228.15699	3	250.83661	3	273.51623	3	296.19586	3	318.87548
4	228.61058	4	251.29020	4	273.96983	4	296.64945	4	319.32907
5	229.06418	5	251.74380	5	274.42342	5	297.10304	5	319.78266
6	229.51777	6	252.19739	6	274.87701	6	297.55663	6	320.23625
7	229.97136	7	252.65098	7	275.33060	7	298.01022	7	320.68985
8	230.42495	8	253.10457	8	275.78420	8	298.46382	8	321.14344
9	230.87855	9	253.55817	9	276.23779	9	298.91741	9	321.59703
510	231.33214	560	254.01176	610	276.69138	660	299.37100	710	322.05062
1	231.78573	1	254.46535	1	277.14497	1	299.82459	1	322.50422
2	232.23932	2	254.91894	2	277.59857	2	300.27819	2	322.95781
3	232.69292	3	255.37254	3	278.05216	3	300.73178	3	323.41140
4	233.14651	4	255.82613	4	278.50575	4	301.18537	4	323.86499
5	233.60010	5	256.27972	5	278.95934	5	301.63896	5	324.31859
6	234.05369	6	256.73331	6	279.41294	6	302.09256	6	324.77218
7	234.50729	7	257.18691	7	279.86653	7	302.54615	7	325.22577
8	234.96088	8	257.64050	8	280.32012	8	302.99974	8	325.67936
9	235.41447	9	258.09409	9	280.77371	9	303.45333	9	326.13296
520	235.86806	570	258.54768	620	281.22731	670	303.90693	720	326.58655
1	236.32165	1	259.00128	1	281.68090	1	304.36052	1	327.04014
2	236.77525	2	259.45487	2	282.13449	2	304.81411	2	327.49373
3	237.22884	3	259.90846	3	282.58808	3	305.26770	3	327.94733
4	237.68243	4	260.36205	4	283.04167	4	305.72130	4	328.40092
5	238.13602	5	260.81565	5	283.49527	5	306.17489	5	328.85451
6	238.58962	6	261.26924	6	283.94886	6	306.62848	6	329.30810
7	239.04321	7	261.72283	7	284.40245	7	307.08207	7	329.76169
8	239.49680	8	262.17642	8	284.85604	8	307.53567	8	330.21529
9	239.95039	9	262.63002	9	285.30964	9	307.98926	9	330.66888
530	240.40399	580	263.08361	630	285.76323	680	308.44285	730	331.12247
1	240.85758	1	263.53720	1	286.21682	1	308.89644	1	331.57606
2	241.31117	2	263.99079	2	286.67041	2	309.35004	2	332.02966
3	241.76476	3	264.44439	3	287.12401	3	309.80363	3	332.48325
4	242.21836	4	264.89798	4	287.57760	4	310.25722	4	332.93684
5	242.67195	5	265.35157	5	288.03119	5	310.71081	5	333.39043
6	243.12554	6	265.80516	6	288.48478	6	311.16441	6	333.84403
7	243.57913	7	266.25876	7	288.93838	7	311.61800	7	334.29762
8	244.03273	8	266.71235	8	289.39197	8	312.07159	8	334.75121
9	244.48632	9	267.16594	9	289.84556	9	312.52518	9	335.20480
540	244.93991	590	267.61953	640	290.29915	690	312.97878	740	335.65840
1	245.39350	1	268.07312	1	290.75275	1	313.43237	1	336.11199
2	245.84710	2	268.52672	2	291.20634	2	313.88596	2	336.56558
3	246.30069	3	268.98031	3	291.65993	3	314.33955	3	337.01917
4	246.75428	4	269.43390	4	292.11352	4	314.79314	4	337.47277
5	247.20787	5	269.88749	5	292.56712	5	315.24674	5	337.92636
6	247.66147	6	270.34109	6	293.02071	6	315.70033	6	338.37995
7	248.11506	7	270.79468	7	293.47430	7	316.15392	7	338.83354
8	248.56865	8	271.24827	8	293.92789	8	316.60751	8	339.28714
9	249.02224	9	271.70186	9	294.38149	9	317.06111	9	339.74073

IX. MASS—AVOIRDUPOIS POUNDS TO KILOGRAMS

[Reduction factor: 1 avoirdupois pound = 0.4535924277 kilogram]

Pounds	Kilos	Pounds	Kilos	Pounds	Kilos	Pounds	Kilos	Pounds	Kilos
750	340.19432	800	362.87394	850	385.55356	900	408.23318	950	430.91281
1	340.64791	1	363.32753	1	386.00716	1	408.68678	1	431.36640
2	341.10151	2	363.78113	2	386.46075	2	409.14037	2	431.81999
3	341.55510	3	364.23472	3	386.91434	3	409.59396	3	432.27358
4	342.00869	4	364.68831	4	387.36793	4	410.04755	4	432.72718
5	342.46228	5	365.14190	5	387.82153	5	410.50115	5	433.18077
6	342.91588	6	365.59550	6	388.27512	6	410.95474	6	433.63436
7	343.36947	7	366.04909	7	388.72871	7	411.40833	7	434.08795
8	343.82306	8	366.50268	8	389.18230	8	411.86192	8	434.54155
9	344.27665	9	366.95627	9	389.63590	9	412.31552	9	434.99514
760	344.73025	810	367.40987	860	390.08949	910	412.76911	960	435.44873
1	345.18384	1	367.86346	1	390.54308	1	413.22270	1	435.90232
2	345.63743	2	368.31705	2	390.99667	2	413.67629	2	436.35592
3	346.09102	3	368.77064	3	391.45027	3	414.12989	3	436.80951
4	346.54461	4	369.22424	4	391.90386	4	414.58348	4	437.26310
5	346.99821	5	369.67783	5	392.35745	5	415.03707	5	437.71669
6	347.45180	6	370.13142	6	392.81104	6	415.49066	6	438.17029
7	347.90539	7	370.58501	7	393.26463	7	415.94426	7	438.62388
8	348.35898	8	371.03861	8	393.71823	8	416.39785	8	439.07747
9	348.81258	9	371.49220	9	394.17182	9	416.85144	9	439.53106
770	349.26617	820	371.94579	870	394.62541	920	417.30503	970	439.98465
1	349.71976	1	372.39938	1	395.07900	1	417.75863	1	440.43825
2	350.17335	2	372.85298	2	395.53260	2	418.21222	2	440.89184
3	350.62695	3	373.30657	3	395.98619	3	418.66581	3	441.34543
4	351.08054	4	373.76016	4	396.43978	4	419.11940	4	441.79902
5	351.53415	5	374.21375	5	396.89337	5	419.57300	5	442.25262
6	351.98772	6	374.66735	6	397.34697	6	420.02659	6	442.70621
7	352.44132	7	375.12094	7	397.80056	7	420.48018	7	443.15980
8	352.89491	8	375.57453	8	398.25415	8	420.93377	8	443.61339
9	353.34850	9	376.02812	9	398.70774	9	421.38737	9	444.06699
780	353.80209	830	376.48171	880	399.16134	930	421.84096	980	444.52058
1	354.25569	1	376.93531	1	399.61493	1	422.29455	1	444.97417
2	354.70928	2	377.38890	2	400.06852	2	422.74814	2	445.42776
3	355.16287	3	377.84249	3	400.52211	3	423.20174	3	445.88136
4	355.61646	4	378.29608	4	400.97571	4	423.65533	4	446.33495
5	356.07006	5	378.74968	5	401.42930	5	424.10892	5	446.78854
6	356.52365	6	379.20327	6	401.88289	6	424.56251	6	447.24213
7	356.97724	7	379.65686	7	402.33648	7	425.01610	7	447.69573
8	357.43083	8	380.11045	8	402.79008	8	425.46970	8	448.14932
9	357.88443	9	380.56405	9	403.24367	9	425.92329	9	448.60291
790	358.33802	840	381.01764	890	403.69726	940	426.37688	990	449.05650
1	358.79161	1	381.47123	1	404.15085	1	426.83047	1	449.51010
2	359.24520	2	381.92482	2	404.60445	2	427.28407	2	449.96369
3	359.69880	3	382.37842	3	405.05804	3	427.73766	3	450.41728
4	360.15239	4	382.83201	4	405.51163	4	428.19125	4	450.87087
5	360.60598	5	383.28560	5	405.96522	5	428.64484	5	451.32447
6	361.05957	6	383.73919	6	406.41882	6	429.09844	6	451.77806
7	361.51316	7	384.19279	7	406.87241	7	429.55203	7	452.23165
8	361.96676	8	384.64638	8	407.32600	8	430.00562	8	452.68524
9	362.42035	9	385.09997	9	407.77959	9	430.45921	9	453.13884

X. METRIC AND ENGLISH EQUIVALENTS OF DISTANCE IN TRACK AND FIELD EVENTS

Metric distances for track and field events to be run in athletic meets held under the jurisdiction of the Amateur Athletic Union were officially adopted by that body on November 22, 1932. The following tables have been included in this book to make it possible for those not familiar with the Metric system to know the various distances expressed in metres.

In Table 1 are given the equivalents of Metric and English distances for principal indoor and outdoor track events.

In Table 2 are given the Metric equivalents for distances in feet, inches and binary fractions of an inch.

The metric equivalent of any distance may be conveniently found, to the nearest ⅛ inch, by breaking the distance down into convenient parts, obtaining the equivalent of each part and then adding them together to get the total equivalent.

For example the metric equivalent of 251 feet, 9½ inches is found as follows:

$$
\begin{array}{rll}
200 \text{ feet} & = 60.960 & \text{metres} \\
50 \text{ feet} & = 15.240 & \text{metres} \\
1 \text{ foot} & = .0305 & \text{metre} \\
9 \text{ inches} & = .229 & \text{metre} \\
½ \text{ inch} & = .013 & \text{metre} \\
\hline
251 \text{ feet, } 9½ \text{ inches} & = 76.4725 & \text{metres}
\end{array}
$$

DISTANCE EQUIVALENTS

Basis $\begin{cases} 1 \text{ metre} = 39.37 \text{ inches} = 3.280\ 8 \text{ feet} = 1.093\ 6 \text{ yards} \\ 1 \text{ kilometre} = 1\ 000 \text{ metres} = 0.621\ 370 \text{ mile} \end{cases}$

TABLE 1.—*Track events*

Yards : Metres		Metres : Yards	
40 =	36. 58	50 =	54. 68
50 =	45. 72	60 =	65. 62
60 =	54. 86	65 =	71. 08
70 =	64. 01	80 =	87. 49
75 =	68. 58	100 =	109. 36
100 =	91. 44	110 =	120. 30
110 =	100. 58	200 =	218. 72
120 =	109. 73	300 =	328. 08
220 =	201. 17	400 =	437. 44
300 =	274. 32	500 =	546. 81
440 =	402. 34 = ¼ mi	600 =	656. 16
600 =	548. 64	800 =	874. 89
880 =	804. 67 = ½ mi	1 000 =	1 093. 61
1 000 =	914. 40	1 500 =	1 640. 42
1 320 = 1 207. 01 = ¾ mi		1 600 =	1 749. 78

Miles : Metres		Metres : Miles		Yards and inches		Miles (approx.)
1 =	1 609. 3	2 000 =	1	427	8	1. 24
2 =	3 218. 7	2 400 =	1	864	24	1. 49
3 =	4 828. 0	3 000 =	1	1 520	30	1. 86
4 =	6 437. 4	3 200 =	1	1 739	20	1. 99
5 =	8 046. 7					
		5 000 =	3	188	2	3. 11
6 =	9 656. 1	6 000 =	3	1 281	24	3. 73
7 =	11 265. 4	10 000 =	6	376	4	6. 21
8 =	12 874. 8	15 000 =	9	564	6	9. 32
9 =	14 484. 1					
		20 000 =	12	752	8	12. 43
10 =	16 093. 5	25 000 =	15	940	10	15. 53
15 =	24 140. 2	30 000 =	18	1 128	12	18. 64
20 =	32 186. 9	50 000 =	31	120	20	31. 07
25 =	40 233. 7					

TABLE 2.—*Field events*

Feet : Metres		Inches : Metres	
1	= 0.305	1	= 0.025
2	= .610	2	= .051
3	= .914	3	= .076
4	= 1.219	4	= .102
5	= 1.524		
		5	= .127
6	= 1.829	6	= .152
7	= 2.134	7	= .178
8	= 2.438	8	= .203
9	= 2.743		
		9	= .229
10	= 3.048	10	= .254
20	= 6.096	11	= .279
30	= 9.144	12	= .305
40	= 12.192		
50	= 15.240	Fractions	
		of an	
60	= 18.288	inch : Metre	
70	= 21.336		
80	= 24.384	⅛	= 0.003
90	= 27.432	¼	= .006
		⅜	= .010
100	= 30.480	½	= .013
200	= 60.960		
300	= 91.440	⅝	= .016
400	= 121.920	¾	= .019
500	= 152.400	⅞	= .022
		1	= .025
600	= 182.880		
700	= 213.360		
800	= 243.840		
900	= 274.321		

The precision of measurement of distance and of time, as ordinarily carried out in track and field events, received consideration in determining the number of decimal places to be carried out in these tables.

Distances in field events are customarily measured in feet and inches to the nearest eighth of an inch. This is about 3 millimetres or 0.0003 metres. In order to convert these measured distances from feet and inches to metres with maximum accuracy, the metric equivalents are given to the nearest 0.001 metre.

The same consideration has been given to the accuracy of measurement of both distance and time as carried out with track events. It should be noted that when time is taken with a stop watch it is usually given to the fifth or tenth of a second. When taken with electrical timing devices it may be given to the hundredth of a second.

In dashes, where 1 second represents a distance of approximately 10 yards or 10 metres, $\frac{1}{10}$ second represents about 1 yard or 1 metre, and $\frac{1}{100}$ second represents a distance of $\frac{1}{10}$ yard or $\frac{1}{10}$ metre. There is no need at present, therefore, to give metric equivalents of distances more precision than the nearest $\frac{1}{10}$ metre, even when the most precise timing methods are used. For distances of less than a mile, however, they have been given to the nearest $\frac{1}{100}$ metre in order to allow for possible future improvement in timing devices.

XI. RECAPITULATION TABLE OF METRIC WEIGHTS AND MEASURES

LINEAR MEASURE

10 millimetres (mm)	= 1 centimetre (cm)
10 centimetres	= 1 decimetre (dm) = 100 millimetres
10 decimetres	= 1 metre (m) = 1 000 millimetres
10 metres	= 1 dekametre (dkm)
10 dekametres	= 1 hectometre (hm) = 100 metres
10 hectometres	= 1 kilometre (km) = 1 000 metres

AREA MEASURE

100 square millimetres (mm^2)	= 1 square centimetre (cm^2)
10 000 square centimetres	= 1 square metre (m^2) = 1 000 000 square millimetres
100 square metres	= 1 are (a)
100 ares	= 1 hectare (ha) = 10 000 square metres
100 hectares	= 1 square kilometre (km^2) = 1 000 000 square metres

VOLUME MEASURE

10 millilitres (ml)	= 1 centilitre (cl)
10 centilitres	= 1 decilitre (dl) = 100 millilitres
10 decilitres	= 1 litre* (l) = 1 000 millilitres
10 litres	= 1 dekalitre (dk$l$)
10 dekalitres	= 1 hectolitre (hl) = 100 litres
10 hectolitres	= 1 kilolitre (kl) = 1 000 litres

CUBIC MEASURE

1 000 cubic millimetres (mm^3)	= 1 cubic centimetre (cm^3)
1 000 cubic centimetres	= 1 cubic decimetre (dm^3) = 1 000 000 cubic millimetres
1 000 cubic decimetres	= 1 cubic metre (m^3) = 1 stere = 1 000 000 cubic centimetres = 1 000 000 000 cubic millimetres

---

* The litre is defined as the volume occupied, under standard conditions, by a quantity of pure water having a mass of 1 kilogram. This volume is very nearly equal to 1 000 cubic centimetres or 1 cubic decimetre; the actual metric equivalent is, 1 litre = 1 000.028 cubic centimetres. (The change in this equivalent from the previously published value of 1 000.027 is based on a recomputation of earlier data, carried out at the International Bureau of Weights and Measures.) Thus the millilitre and the litre are larger than the cubic centimetre and the cubic decimetre, respectively, by 28 parts in 1 000 000; except for determinations of high precision, this difference is so small as to be of no consequence.

MASS

10 milligrams (mg) = 1 centigram (cg)
10 centigrams = 1 decigram (dg) = 100 milligrams
10 decigrams = 1 gram (g) = 1 000 milligrams
10 grams = 1 dekagram (dkg)
10 dekagrams = 1 hectogram (hg) = 100 grams
10 hectograms = 1 kilogram (kg) = 1 000 grams
1 000 kilograms = 1 metric ton (t)

NOTE.—In the metric system of weights and measures, designations of multiples and sub-divisions of any unit may be arrived at by combining with the name of the unit the prefixes *deka*, *hecto*, and *kilo*, meaning, respectively, 10, 100, and 1 000, and *deci*, *centi*, and *milli*, meaning, respectively, one-tenth, one-hundredth, and one-thousandth. In some of the fore-going metric tables, some such multiples and subdivisions have not been included for the reason that these have little, if any, currency in actual usage.

In certain cases, particularly in scientific usage, it becomes convenient to provide for multi-ples larger than 1 000 and for subdivisions smaller than one-thousandth. Accordingly, the following prefixes have been introduced and these are now generally recognized.

myria, meaning 10 000
mega, meaning 1 000 000
micro, meaning one-millionth

A special case is found in the term "micron" (abbreviated as μ [the Greek letter m u]), a coined word meaning one-millionth of a metre (equivalent to one-thousandth of a milli-metre); a millimicron (abbreviated as m μ) is one-thousandth of a micron (equivalent to one-millionth of a millimetre), and a micromicron (abbreviated as μμ) is one-millionth of a micron (equivalent to one-thousandth of a millimicron or to 0.000 000 001 millimetre.)

XII. TABLES OF UNITED STATES CUSTOMARY WEIGHTS AND MEASURES

LINEAR MEASURE

12 inches (in.)	= 1 foot (ft)
3 feet	= 1 yard (yd)
5½ yards	= 1 rod (rd), pole, or perch = 16½ feet
40 rods	= 1 furlong (fur.) = 220 yards = 660 feet
8 furlongs	= 1 statute mile (mi) = 1 760 yards = 5 280 feet
3 miles	= 1 league = 5 280 yards = 15 840 feet
6 080.20 feet	= 1 nautical, geographical, or sea mile

AREA MEASURE*

144 square inches (sq in.)	= 1 square foot (sq ft)
9 square feet	= 1 square yard (sq yd) = 1 296 square inches
30¼ square yards	= 1 square rod (sq rd) = 272¼ square feet
160 square rods	= 1 acre = 4 840 square yards = 43 560 square feet
640 acres	= 1 square mile (sq mi)
1 mile square	= 1 section [of land]
6 miles square	= 1 township = 36 sections = 36 square miles

CUBIC MEASURE*

1 728 cu inches	= 1 cu foot
27 cu feet	= 1 cu yard
128 cu feet	= 1 cord of wood
144 cu inches	= 1 board foot (volume of board 1 foot square by 1 inch thick)

GUNTER'S OR SURVEYORS CHAIN MEASURE

7.92 inches (in.)	= 1 link (li)
100 links	= 1 chain (ch) = 4 rods = 66 feet
80 chains	= 1 statute mile (mi) = 320 rods = 5 280 feet

---

* Squares and cubes of units are sometimes abbreviated by using "superior" figures. For example, ft² means square foot, and ft³ means cubic foot.

LIQUID MEASURE*

4 gills	= 1 pint
2 pints	= 1 quart
4 quarts	= 1 gallon
7.48 gallons	= 1 cu foot
240 gallons of water	= 1 ton
340 gallons of gasoline	= 1 ton

APOTHECARIES FLUID MEASURE

60 minims (min) = 1 fluid dram (fl dr) [= 0.225 6 cubic inch]
8 fluid drams = 1 fluid ounce (fl oz) [= 1.804 7 cubic inches]
16 fluid ounces = 1 pint (pt) [= 28.875 cubic inches] = 128 fluid drams
2 pints = 1 quart (qt) [= 57.75 cubic inches] = 32 fluid ounces = 256 fluid drams
4 quarts = 1 gallon (gal) [= 231 cubic inches] = 128 fluid ounces = 1 024 fluid drams

DRY MEASURE**

2 pints (pt) = 1 quart (qt) [= 67.200 6 cubic inches]
8 quarts = 1 peck (pk) [= 537.605 cubic inches] = 16 pints
4 pecks = 1 bushel (bu) [= 2 150.42 cubic inches] = 32 quarts

AVOIRDUPOIS MASS***

[The "grain" is the same in avoirdupois, troy, and apothecaries mass.]

$27\frac{11}{32}$ grains = 1 dram (dr)
16 drams = 1 ounce (oz) = $437\frac{1}{2}$ grains
16 ounces = 1 pound (lb) = 256 drams = 7 000 grains
100 pounds = 1 hundredweight (cwt)****
20 hundredweights = 1 ton (tn) = 2 000 pounds****
In "gross" or "long" measure, the following values are recognized:
112 pounds = 1 gross or long hundredweight****
20 gross or long
    hundredweights = 1 gross or long ton = 2 240 pounds****

---

* When necessary to distinguish the *liquid* pint or quart from the *dry* pint or quart, the word "liquid" or the abbreviation "liq" should be used in combination with the name or abbreviation of the *liquid* unit.
** When necessary to distinguish the *dry* pint or quart from the *liquid* pint or quart, the word "dry" should be used in combination with the name or abbreviation of the dry unit.
*** When necessary to distinguish the *avoirdupois* dram from the *apothecaries* dram, or to distinguish the *avoirdupois* dram or ounce from the *fluid* dram or ounce, or to distinguish the *avoirdupois* onnce or pound from the *troy* or *apothecaries* ounce or pound, the word "avoirdupois" or the abbreviation "avdp" should be used in combination with the name or abbreviation of the *avoirdupois* unit.
**** When the terms "hundredweight" and "ton" are used unmodified, they are commonly understood to mean the 100-pound hundredweight and the 2 000-pound ton, respectively, these units may be designated "net" or "short" when necessary to distinguish them from the corresponding units in *gross* or *long* measure.

161

TROY MASS

[The "grain" is the same in avoirdupois, troy, and apothecaries mass.]

24 grains   = 1 pennyweight (dwt)
20 pennyweights = 1 ounce troy (oz t) = 480 grains
12 ounces troy   = 1 pound troy (lb t) = 240 pennyweights =
                   5 760 grains

APOTHECARIES MASS

[The "grain" is the same in avoirdupois, troy, and apothecaries mass.]

20 grains   = 1 scruple (s ap)
 3 scruples   = 1 dram apothecaries (dr ap) = 60 grains
 8 drams apothecaries = 1 ounce apothecaries (oz ap) = 24
                          scruples = 480 grains
12 ounces apothecaries = 1 pound apothecaries (lb ap) = 96
                          drams apothecaries = 288 scruples =
                          5 760 grains

CIRCULAR MEASURE

 60 seconds = 1 minute
 60 minutes = 1 degree
 60 degrees = 1 quadrant
360 degrees = 1 circumference

162

XIII. ALPHABETICAL CONVERSION TABLE

The fundamental purpose of this chart is to furnish a source of reference for units, standards, and conversion factors

1 acre $\begin{cases} =160 \text{ square rods.} \\ =4,840 \text{ square yards.} \\ =43,560 \text{ square feet.} \end{cases}$

1 barrel = 7,056 cubic inches.

1 board foot $\begin{cases} =144 \text{ cubic inches.} \\ =2,360 \text{ cubic centimetres.} \end{cases}$

1 B. t. u. (British thermal unit) $\begin{cases} =778 \text{ foot pounds.} \\ =0.2930 \text{ international watt hour.} \\ =0.252 \text{ calorie (I. T.).} \end{cases}$

1 bushel $\begin{cases} =2,150.42 \text{ cubic inches.} \\ =1\frac{1}{4} \text{ cubic feet, approx.} \end{cases}$

1 calorie (I. T.) $\begin{cases} =1/860 \text{ international watt hours.} \\ =3.97 \times 10^{-3} \text{ B. t. u.} \end{cases}$

1 carat, metric $\begin{cases} =200 \text{ metric milligrams.} \\ =3.0865 \text{ grains.} \end{cases}$

1 centare (square metre) $\begin{cases} =10.764 \text{ square feet.} \\ =1.196 \text{ square yards.} \end{cases}$

1 centimetre = 0.3937 inch.

1 chain (engineers) $\begin{cases} =100 \text{ links of 1 foot each.} \\ =30.48 \text{ metres.} \end{cases}$

1 chain (surveyors or Gunters) $\begin{cases} =4 \text{ rods.} \\ =100 \text{ links.} \\ =66 \text{ feet.} \\ =20.1 \text{ metres.} \end{cases}$

1 cheval (French horsepower) = 0.986 horsepower.

1 circular mil $\begin{cases} =\text{Area of circle whose diameter is 1 mil, or} \\ \quad 1/1000 \text{ inch.} \\ =0.000000785 \text{ square inch.} \end{cases}$

1 cord $\begin{cases} =128 \text{ cubic feet.} \\ =3.625 \text{ cubic metres.} \end{cases}$

1 cubic foot $\begin{cases} =1,728 \text{ cubic inches.} \\ =60 \text{ pints.} \\ =0.8 \text{ bushel.} \\ =1,000 \text{ ounces of water, approx.} \\ =0.028 \text{ cubic metre.} \\ =28.32 \text{ litres.} \end{cases}$

1 cubic foot of water $\begin{cases} =62.4 \text{ pounds.} \\ =1,000 \text{ ounces, approx.} \end{cases}$

1 cubic inch = 16.39 cubic centimetres.

1 cubic metre $\begin{cases} =35.314 \text{ cubic feet.} \\ =1.308 \text{ cubic yards.} \end{cases}$

1 cubic yard $\begin{cases} =27 \text{ cubic feet.} \\ =0.765 \text{ cubic metre.} \end{cases}$

1 decimetre = 3.937 inches.

1 dram (fluid) $\begin{cases} =60 \text{ minims.} \\ =3.697 \text{ millilitres.} \\ =4 \text{ cubic centimetres, approx.} \end{cases}$

1 em, 1 pica (printing industry)=1/6 of an inch.

1 fathom (nautical) $\begin{cases} =6 \text{ feet.} \\ =1.83 \text{ metres.} \end{cases}$

1 fluid ounce $\begin{cases} =8 \text{ fluid drams.} \\ =29.573 \text{ millilitres.} \end{cases}$

1 foot $\begin{cases} =12 \text{ inches.} \\ =0.305 \text{ metre.} \end{cases}$

1 foot pound=0.1383 kilogrammetre.

1 furlong (British) $\begin{cases} =220 \text{ yards.} \\ =201.2 \text{ metres.} \end{cases}$

1 gallon (U.S.) $\begin{cases} =231 \text{ cubic inches.} \\ =4 \text{ quarts.} \\ =8 \text{ pints.} \\ =3.875 \text{ litres.} \\ =128 \text{ fluid ounces.} \end{cases}$

1 gallon of water=8.33 pounds at 62° F. (16.67° C.) in air.

1 gallon per cubic foot=133.7 litres per cubic metre.

Gallon (British Imperial and Canadian). $\begin{cases} =277.4 \text{ cubic inches.} \\ =1.201 \text{ U.S. gallons.} \\ =\text{volume of 10 pounds water at 62°} \\ \quad \text{F. (16.67° C.).} \\ =4.546 \text{ litres.} \end{cases}$

1 gill=¼ pint.

1 grain $\begin{cases} =1/7000 \text{ pound avoirdupois.} \\ =0.0648 \text{ gram.} \end{cases}$

1 gram $\begin{cases} =15.43 \text{ grains.} \\ =0.0353 \text{ ounce.} \\ =0.0022 \text{ pound.} \end{cases}$

1 hand=4 inches.

1 hectare (square hectometre)=2.47 acres.

1 horsepower $\begin{cases} =33,000 \text{ foot-pounds per minute.} \\ =42.41 \text{ B. t. u. per minute.} \\ =1.014 \text{ chevals.} \\ =746 \text{ watts.} \end{cases}$

1 hundredweight (British) $\begin{cases} =112 \text{ pounds.} \\ =50.80 \text{ kilograms.} \end{cases}$

1 inch=25.4 millimetres.

1 kilogram $\begin{cases} =2.2046 \text{ pounds.} \\ =35.274 \text{ ounces.} \\ =15432.36 \text{ grains.} \\ =0.0011 \text{ short ton.} \\ =0.00098 \text{ long ton.} \end{cases}$

1 kilometre $\begin{cases} =1000 \text{ metres.} \\ =0.621 \text{ mile.} \end{cases}$

1 kilowatt $\begin{cases} =1.34 \text{ horsepower.} \\ =56.9 \text{ B. t. u. per minute.} \end{cases}$

1 knot (nautical, speed) $\begin{cases} =6080.20 \text{ feet per hour.} \\ =1.85 \text{ kilometres per hour.} \end{cases}$

1 light year $\begin{cases} =5.9\times10^{12} \text{ miles.} \\ =9.5\times10^{12} \text{ kilometres.} \end{cases}$

1 link (surveyors measure) $\begin{cases} =0.66 \text{ foot.} \\ =0.201 \text{ metre.} \end{cases}$

1 litre $\begin{cases} =1.000028 \text{ cubic decimetres.} \\ =0.264 \text{ gallon.} \\ =1.057 \text{ quarts.} \\ =61.03 \text{ cubic inches.} \\ =0.035 \text{ cubic feet.} \\ =33.8148 \text{ fluid ounces.} \\ =270.518 \text{ fluid drams.} \end{cases}$

1 metre $\begin{cases} =39.37 \text{ inches.} \\ =3.28 \text{ feet.} \\ =1.09 \text{ yards.} \\ =1\ 650\ 763.73 \text{ wave lengths in a vacuum of} \\ \quad \text{orange-red radiation of krypton 86.} \end{cases}$

1 metric ton $\begin{cases} =2204.6 \text{ pounds.} \\ =1.1023 \text{ short tons.} \end{cases}$

1 microgram=1/1000 milligram.

1 mil $\begin{cases} =0.001 \text{ inch.} \\ =25.4 \text{ microns.} \\ =0.0254 \text{ millimetre.} \end{cases}$

1 mile $\begin{cases} =1760 \text{ yards.} \\ =5280 \text{ feet.} \\ =320 \text{ rods.} \\ =1.61 \text{ kilometres.} \end{cases}$

1 milligram=0.0154 grain.

1 millilitre (see litre above) $\begin{cases} =1.000028 \text{ cubic centimetres.} \\ =0.0610 \text{ cubic inch.} \end{cases}$

1 minim (fluid) $\begin{cases} =1/60 \text{ fluid dram.} \\ =1/480 \text{ fluid ounce.} \end{cases}$

1 ounce (avoirdupois, ordinary) $\begin{cases} =437.5 \text{ grains.} \\ =0.911 \text{ troy ounce.} \\ =0.0000279 \text{ long ton.} \\ =28.35 \text{ grams.} \end{cases}$

1 ounce, fluid $\begin{cases} =1.805 \text{ cubic inches.} \\ =29.573 \text{ millilitres.} \end{cases}$

1 ounce, troy $\begin{cases} =480 \text{ grains.} \\ =31.103 \text{ grams.} \end{cases}$

1 perch (British) $\begin{cases} =30.25 \text{ square yards.} \\ =1/160 \text{ acre.} \end{cases}$

1 pied (French foot) $\begin{cases} =12 \text{ Paris inches.} \\ =0.325 \text{ metre.} \end{cases}$

1 pint=0.4732 litre.

1 point (printers type)=1/72 inch.

1 pole (British) $\begin{cases} =5\frac{1}{2} \text{ yards.} \\ =5.03 \text{ metre.} \\ =1 \text{ rod.} \end{cases}$

165

1 pouce (Paris inch)=2.71 centimetre.

1 pound (avoirdupois, ordinary) $\begin{cases} =16 \text{ ounces.} \\ =7000 \text{ grains.} \\ =454 \text{ grams.} \\ =0.454 \text{ kilogram.} \\ =14.58 \text{ troy ounces.} \end{cases}$

1 pound per cubic foot=16.02 kilogram per cubic metre.
1 pound per square inch=0.433×head of water (in feet).
1 pound per square inch=0.0703 kilogram per square centimetre.
1 pound per square foot=4.88 kilogram per square metre.

1 quart $\begin{cases} =2 \text{ pints.} \\ =\frac{1}{4} \text{ gallon.} \\ =0.946 \text{ litre.} \end{cases}$

1 quarter (British quarter hundredweight) $\begin{cases} =28 \text{ pounds.} \\ =12.70 \text{ kilograms.} \end{cases}$

1 rod (surveyor's measure) $\begin{cases} =16.5 \text{ feet.} \\ =25 \text{ links.} \\ =5.03 \text{ metres.} \end{cases}$

1 rood (British) $\begin{cases} =40 \text{ perches.} \\ =\frac{1}{4} \text{ acre.} \end{cases}$

1 square centimetre=0.155 square inch.
1 square foot=0.093 square metre.
1 square inch=6.452 square centimetres.
1 square kilometre=0.386 square mile.

1 square metre (centare) $\begin{cases} =10.764 \text{ square feet.} \\ =1.196 \text{ square yards.} \end{cases}$

1 square mil $\begin{cases} =0.000001 \text{ square inch.} \\ =0.00000645 \text{ square centimetre.} \end{cases}$

1 square mile $\begin{cases} =640 \text{ acres.} \\ =3,097,600 \text{ square yards.} \\ =2.59 \text{ square kilometres.} \end{cases}$

1 square millimetre=0.00155 square inch.
1 square rod=25.29 square metres.
1 square yard=0.836 square metre.

1 stone (British) $\begin{cases} =14 \text{ pounds.} \\ =6.35 \text{ kilograms.} \end{cases}$

1 ton (short) $\begin{cases} =2,000 \text{ pounds.} \\ =907 \text{ kilograms.} \end{cases}$

1 ton (long) $\begin{cases} =2,240 \text{ pounds.} \\ =1,016 \text{ kilograms.} \end{cases}$

1 ton (metric) $\begin{cases} =1,000 \text{ kilograms.} \\ =2,204.62 \text{ pounds.} \end{cases}$

1 yard $\begin{cases} =3 \text{ feet.} \\ =36 \text{ inches.} \\ =0.914 \text{ metre.} \\ =1\ 508\ 798.05 \text{ wave lengths in a vacuum of} \\ \quad \text{orange-red radiation of krypton 86.} \end{cases}$

Index

XV. ANSWERS TO PRACTICE EXAMPLES

Page 34

1. (a) 10 mm; 100 mm; 1,000 mm
   (b) 10 cm; 100 cm
   (c) 10 dm; 7.5 dm; 105 dm
2. (a) 3.5 m (b) 80 mm (c) 3.25 m
   (d) 7,500 mm (e) 82.5 mm (f) 3.75 m
   (g) 3.75 m (h) 1,750 mm (i) 3.25 m
   (j) 20 m
3. (a) 1.765 m (b) 0.925 m (c) 12.012 m
4. (a) 9.700 mm (b) 727.5 cm
   (c) 333.75 dm

Page 35

7. (a) 35 cm (b) 5,000 mm (c) 100 mm
   (d) 10 dm (e) 25 dm (f) 50 cm
8. (a) 60.08 m (b) 9.79 m (c) 7.54 m
   (d) 18,000.00 m or 18.00 km
10. 4.55 m

Page 36

1. (a) 1,000 mg (b) 100 cg (c) 10 dg
   (d) 1,000 g
2. (a) 50 cg (b) 22.5 dg (c) 6.5 g
   (d) 3,000 mg (e) 400 cg (f) 50 dg
3. (a) 80 cl (b) 70 cl (c) 50 ml (d) 90 cl
   (e) 390 ml (f) 37 ml

Page 38

1. (1) 1068.8 km (2) 1904 km
   (3) 1609.6 km (4) 1878.4 km
   (5) 1379.2 km (6) 2121.6 km
   (7) 1212.8 km (8) 1768 km (9) 1264 km
   (10) 1324.8 km

Page 39

(1) 90 m (2) 130.5 m (3) 36 m
(4) 73.8 m (5) 22.95 m (6) 1.20 m
(7) 18 m; 36 m (8) 20.85 m (9) 65.4 m
(10) 2.25 m

169

Page 40

(1) 195 cm (2) 900 cm (3) 360 cm
(4) 420 cm (5) 2.700 cm (6) 75 cm
(7) 15 cm (8) 105 cm; 130 cm (9) 270 cm
(10) 22.5 cm

Page 41

(6) 8.125 cm; 21.875 cm (8) 187.5 cm
(9) 30 cm × 20 cm × 15 cm
(10) 51.25 cm

Page 42

(1) 225 mm; 300 mm (3) 118.75 mm
(4) 37.5 mm (5) 12.5 mm (8) 550 mm
(9) 37.5 mm (10) 181.25 mm

Page 42

(1) 85 cm; 63.75 cm; 90 cm; 41.25 cm;
2.6 m (2) 90 cm; 67.5 cm; 95 cm;
41.25 cm; 2.6 m

Page 43

(3) 95 cm; 72.5 cm; 100 cm; 41.875 cm;
2.6 m (4) 100 cm; 77.5 cm; 105 cm;
42.5 cm; 2.6 m (5) 110 cm; 90 cm;
115 cm; 43.4 cm; 2.7 m (6) 115 cm;
95 cm; 120 cm; 43.75 cm; 2.9 m

Page 44

(1) 2 oz. (2) 1½ oz. (3) 1¼ oz.
(4) 1⅛ oz. (5) 2¼ oz. (6) 1 lb.
(7) 2½ lbs. (8) 9½ oz. (9) 7½ oz.
(10) 8 oz. (11) 198 kg (12) 173 kg
(13) 149 kg (14) 297 kg (15) 495 kg

Page 44

(1) 454 g (2) 908 g (3) 2.25 kg

Page 45

(4) 4.5 kg (5) 1.24 kg (6) 1.58 kg
(7) 112 g (8) 227 g (9) 1.90 l
(10) 0.95 l (11) 1.43 l (12) 3.8 l
(13) 4.5 m (14) 3.15 m (15) 13.5 m
(1) 500 g (2) 50 g (3) 78.875 kg
(4) 200.75 kg (5) 3.585 kg; 21.415 kg

Page 46

(6) 3,000 ml; 300 cl (7) 10 loaves
(8) 3.55 kg (9) 15 cm (10) 40 mph;
50 mph; 60 mph (12) 176.6 °C